Cost Analysis and Estimating

Willis R. Greer, Jr. Daniel A. Nussbaum
Editors

Cost Analysis and Estimating

Tools and Techniques

With 48 Figures

Springer Science+Business Media, LLC

Willis R. Greer, Jr.
College of Business Administration
University of Iowa
Iowa City, IA 52242
USA

Daniel A. Nussbaum
Naval Center for Cost Analysis
Washington, D.C. 20350-1100
USA

Library of Congress Cataloging-in-Publication Data
Greer, Willis R.
Cost analysis and estimating : tools and techniques / Willis R. Greer, Jr., Daniel Nussbaum.
p. cm.
1. Costs, Industrial—Estimates. 2. Manufacturing processes--Costs—Estimates. 3. Manufactures—Costs—Estimates. I. Nussbaum, Daniel, 1943– II. Title.
TS167.G74 1990
658.15'52—dc20 90-9783

Printed on acid-free paper.

Originally published by Springer-Verlag New York, Inc. in 1990
Softcover reprint of the hardcover 1st edition 1990

Camera-ready text prepared by the authors.

9 8 7 6 5 4 3 2 1

ISBN 978-1-4612-6976-2 ISBN 978-1-4612-0995-9 (eBook)
DOI 10.1007/978-1-4612-0995-9

FOREWORD

Changes in production processes reflect the technological advances permeating our products and services. U.S. industry is modernizing and automating. In parallel, direct labor is fading as the primary cost driver while engineering and technology related cost elements loom ever larger. Traditional, labor-based approaches to estimating costs are losing their relevance. Old methods require augmentation with new estimating tools and techniques that capture the emerging environment. This volume represents one of many responses to this challenge by the cost analysis profession.

The Institute of Cost Analysis (ICA) is dedicated to improving the effectiveness of cost and price analysis and enhancing the professional competence of its members. We encourage and promote exchange of research findings and applications between the academic community and cost professionals in industry and government. The 1990 National Meeting in Los Angeles, jointly sponsored by ICA and the National Estimating Society (NES), provides such a forum. Presentations will focus on new and improved tools and techniques of cost analysis.

This volume is the second in a series. The first was produced in conjunction with the 1989 National Meeting of ICA/NES in Washington, D.C. The articles in this volume, all refereed, were selected from about 100 submitted for presentation at the Los Angeles meeting.

On behalf of the cost professionals who will benefit from this volume, I want to thank those who brought it to us. Professor Willis R. Greer, Jr. and Dr. Daniel A. Nussbaum, the editors, dealt ably with an extremely tight schedule and gained commendable cooperation from the authors. Once again, we thank Professor Thomas R. Gulledge who conceived the idea for the series and worked with the editors and authors to perpetuate it.

Stephen J. Balut
President
Institute of Cost Analysis

PREFACE

The articles which appear in this volume could have been published in a variety of high quality, scholarly journals. Among them would have been operations research, information systems, economics, and defense systems journals.

However, these authors chose to make these manuscripts available to us rather than the more conventional literature for laudable purpose. They recognized that the readers of this volume, the attendees of the ICA/NES Conference, and others, would benefit by having their work assembled in one convenient reference manual. Accordingly, the reader is treated to a broad cross section of work with a single focus; to improve our ability to estimate and analyze the cost of defense. We, the editors, and the readers are deeply indebted to the authors, whose selfless contributions made this volume possible.

We are also deeply indebted to the following individuals, who served as referees for the complex task of reviewing manuscripts, some of which have been published here and others not.

Steve Balut
Avijit Banerjee
Peter Beck
Dan C. Boger
Steve Book
J. D. Camm
Sidhartha R. Das
Ed Deane
Brian Flynn
Thurman Gardner
Thomas R. Gulledge
John Honig
Roland Kankey
David Lamm
Schuyler C. Lawrence
Bin-Shan Lin
Lewis A. Litteral
Bruce Miller
R. P. Mohanty
Doug Moses
Richard Nelson
Michael Peters
Dan Sheldon
Mike Sovereign
Philip Tsung
Karen Tyson
Rsit Unal
V. Valdmanis
H.D. Vinod
Ted Wallenius
James Weathersbee
J.C. Westland
Norman K. Womer

CONTENTS

V. Quality Control Techniques

VI. Warranties and Repair Parts Costing

VII. Test and Evaluation Issues

I. Cost Estimating and Changing Technology

REESTIMATING THE COST OF PRODUCTION IN A FUZZY TECHNOLOGICAL ENVIRONMENT

BY
H. D. VINOD and PARANTAP BASU
Economics Dept, Fordham University, New York, 10458

ABSTRACT

Most profit maximizing firms are aware that the boundary of the feasible set of technological opportunities is fuzzy. The neoclassical theory of the firm assumes that the boundary of the feasible set of technological opportunities is well-defined. We use a spheroidal neighborhood of a neoclassical profit maximizing equilibrium as a suitable approximation of a fuzzy input demand correspondence. Using trigonometric functions this spheroidal neighborhood is parameterized and a fuzzy cost function is generated. We use Nerlove's (1963) famous electric utility data to estimate a generalized Leontief cost function, with and without inclusion of a fuzz factor. The concept of fuzz used here seems to be in accord with Popper's (1973) notion of "plastic control" and Simon's notion of "bounded rationality". For concreteness we plot a fuzzy isoquant for Nerlove's data. We also report numerical estimates of scale elasticity, marginal cost, Hessian matrix, etc.

1. Introduction

The traditional neoclassical theory of the firm generally accepts that firms maximize their objective function in a precisely defined technological environment. It is usually assumed that the producer knows without any vagueness the boundary of his feasible set of technological opportunities. Technological uncertainty is treated in terms of the producer's rational expectations about the objective probability distribution of the random shock driving the technology. Actually, all producing agents cannot be expected to be well versed in probability theory when they make their production and sales decisions. It is now recognized in the literature that the observed behavior of agents can be imprecise or fuzzy. The fundamental premise of fuzzy set theory arises from a value-loaded description of the environment. According to the Paretian notion, technological alternatives are either "efficient," meaning that they are on the boundary of the feasible set, or "inefficient," which means that they are inside the production possibility set. In practice, a company manager is likely to perceive varying degrees of technological efficiency (e.g., very efficient, less efficient, not so efficient etc.). In this context, an application of fuzzy set theory is potentially useful.

In the literature, this issue has been slowly becoming prominent. Lester (1947) drew attention to the inadequacy of marginalism to deal with this kind of situation and argued that the equilibrium is a vague zone rather than a well defined point. Hirshleifer and Riley (1979, 1981) recognize that this type of fuzzy uncertainty about an agent's environment may have possible implications for his decision. There are several attempts to formalize the fuzzy nature of a producer's behavior (see, Blin et al.(1974), Chang (1977), Ponsard (1982), Taranu (1977)). Chen and Yu (1988) derive some useful properties of the fuzzy production correspondence and show how the deterministic analysis of production generalizes in a fuzzy environment. Following Savage's focal decision theory, McCain (1987) introduces a notion of fuzzy confidence interval and applies it to price theory. There may be other attempts to formalize theoretically a fuzzy production environment in the vast literature on the subject. The measurement of fuzz in technology is important because it has definite implications for the estimation of a firm's cost function.

In this paper, we suggest a new methodology for estimating cost and input demand functions in a technological environment with fuzz. We approximate an input correspondence with fuzz in terms of an ellipsoidal neighborhood around a standard neoclassical profit maximizing equilibrium. Using trigonometric functions we first obtain a parametric form for the input correspondence. In the next step, we derive a firm's cost function using these parametric forms. Applying the duality theory, we illustrate that this cost function with fuzz can be used to generate a fuzz technology of empirical interest. Using Nerlove's (1963) famous electric utility data, we finally estimate a generalized Leontief cost function incorporating the fuzz factor. As one might expect, it turns out that the neoclassical non-fuzzy (we call it crisp) equilibrium is a special case of our cost function. Statistical tests are conducted to verify the observed importance of the proposed fuzz in the technology.

The paper is organized as follows. In Section 2, we review

the notion of a fuzzy membership function in a general context. This idea is applied to a standard production model and an approximation is suggested to identify the fuzzy set of input choices for given factor prices. We illustrate graphically how the standard duality theorem can be modified to map the input correspondence to the fuzzy technology. With our formulation in Section 3 we estimate a cost function using Nerlove's (1963) electric utility data and quantify the fuzz factor in the cost and technology facing the firm.

2. Theory of Production with Fuzzy Set

In this section, we review some basic concepts from the fuzzy set theory and indicate their implications for production theory. Fuzziness arises from a lack of well defined boundaries. Let M be a subset of the universe U. The set M is said to be fuzzy if the transition between membership and nonmembership of M is gradual not abrupt. If the transition is abrupt, the set is called crisp (not fuzzy). In this case an element is either a full member of the set M or it is a nonmember. This binary situation may be summarized by a set characteristic function, μ(e, M) is defined as:

$$\mu\,(e, M) \quad = \begin{cases} 1 & \text{if } e \in M \\ 0 & \text{if } e \notin M \end{cases} \tag{2.1}$$

If the set M is not crisp, there are members of M which belong to either class, having a membership rank μ somewhere between 0 and 1. This notion can be formalized by a membership function following Zadeh (1965). A membership function, μ^*(e,M) is defined as

$$\mu^*(e,M)\colon M \to [0,1] \tag{2.2}$$

Note that set characteristic function μ of (2.1) is a special case of the membership function μ^*, arising when e can be classified clearly into one of two slots, either a full member or no member of the set M. Observe the following contrasts with the ordinary set theory. When M is a fuzzy set, $M \cup M^c$, where M^c denotes its complement, is not equal to the universe U. Furthermore, $M \cap M^c$ is not the null set. Kotz and Johnson (1988) and Vinod and Basu(1989) provide further details and references regarding fuzzy set theory.

Let Ω denote a set of n input prices and x denote an $n \times 1$ vector of corresponding input demand quantities. The demand correspondence is defined as a mapping from Ω to

$\Re^n$, the n-dimensional space of real numbers. Let the image of x in $\Re^n$ be denoted by $x(\Omega)$. When x is crisp, the image $x(\Omega)$ is also crisp. On the other hand, when x is fuzzy so is $x(\Omega)$. Our task is to approximate the fuzzy set $x(\Omega)$ for empirical implementation [1]. [See End notes before references]
Ideally the set $x(\Omega)$ should be derived using some fuzzy optimization procedure. This is an ambitious and difficult job, given the present state of fuzzy set optimization in economics. It is also questionable how suitable such a procedure will be for empirical purposes. We follow a less ambitious strategy, which is empirically tractable. Let us start with the profit maximizing input choice denoted by x^*. A suitable approximation of $x(\Omega)$ in a fuzzy production environment is a neighborhood around this x^*. We assume that $x(\Omega)$ remains on the surface of a closed ellipsoid or spheroid $S(x^*)$ centered at the profit maximizing (tangency) point x^* with a parametric (empirically determined) radius. Figure 1 provides an intuitive illustration in a two dimensional input plane. Now, x^* is shown as the tangency point between the isocost line and a crisp isoquant $I(y_0)$ for the output level y_0. When technology is fuzzy, the isoquant corresponding to the level of output y_0 becomes a thick frontier, shown by the dotted curves. The optimal input choice will be somewhere in the neighborhood of x^*. A unit circle is drawn with x^* at the center to represent the possible choice of inputs in this case. The choice of input x_1 is given by the fuzzy interval $\tilde{x}_1$, which is a projection of the this circle on the x_1 axis. Similarly, $\tilde{x}_2$ is a projection of this circle on the x_2 axis. The circle represents the fuzzy region of input quantities of x_1 and x_2 demanded by the firm. This circle becomes an ellipsoid in a more general setting. The reason we choose this specification is, that then the fuzzy input demand functions can be parameterized by the following trigonometric functions, which lead to considerable analytical convenience. It is straightforward to verify that in this two dimensional case:

$$\tilde{x}_1 = x_1^* + r \cos\theta \tag{2.3}$$

$$\tilde{x}_2 = x_2^* + r \sin\theta \tag{2.4}$$

Note that θ is a parameter which represents the fuzz in spherical coordinates. This formulation gives rise to a wide set of possibilities. If θ equals zero, $\tilde{x}_2 = x_2^*$ which means that the optimal choice of x_2 is the same in both fuzzy and crispy cases. Since cosine of zero is unity, the entire fuzz is reflected in the first factor demand. All input demands are not crisp even when $\theta=0$, because $\cos 0=1$ in (2.3). Similarly, for θ equal to $\pi/2$, there is only fuzz in $\tilde{x}_2$ and the fuzz in $\tilde{x}_1$ disappears. For the θ value between 0 and 2π, we have an interesting set of possibilities. Of course, the actual value of θ is an empirical question. Since $-1\leq\sin\theta\leq 1$ and $-1\leq\cos\theta\leq 1$, it is obvious that $x_i^* - r\leq \tilde{x}_i\leq x_i^* + r$ represents the ex ante fuzzy region in which the fuzzy input demands $\tilde{x}_i$ must lie for i=1,2 for any value of θ. This two-dimensional region is depicted in Figure 1. After the inputs are chosen, θ and $\tilde{x}_i$ are fixed ex post.

For a three dimensional input space, we consider an ellipsoid of the following kind:

$$(\tilde{x}_1 - x_1^*)^2/r_1^2 + (\tilde{x}_2 - x_2^*)^2/r_2^2 + (\tilde{x}_3 - x_3^*)^2/r_3^2 = 1 \qquad (2.5)$$

For convenience we use $r_1=r_2=r_3=r$ throughout this paper, implying a sphere instead of an ellipsoid. However, ellipsoidal regions can be considered by using different r_i values. The fuzzy input demand functions, $\tilde{x}_i$ can be parameterized by sine and cosine functions as follows.

$$\tilde{x}_1 = x_1^* + r\cos\theta \qquad (2.6)$$

$$\tilde{x}_2 = x_2^* + r\sin\theta\cos\phi \qquad (2.7)$$

$$\tilde{x}_3 = x_3^* + r\sin\theta\sin\phi \qquad (2.8)$$

The formulation involving trigonometric functions can be easily generalized to n (n>3) dimensions; in which case we have a spheroid. An Appendix explains the use of trigonometric functions to obtain explicit expression for coordinates when n>3.

It is interesting to observe that we need only one angle, θ to express the case of two dimensions in (2.3) and (2.4). With three dimensions, we need two angles, θ and ϕ in (2.6) to (2.8). For n dimensions we need n-1 angular parameters. Evidently these trigonometric functions impose certain econometric identifying restrictions on the observable cost function. Our methods permit explicit derivation of these identifying restrictions.

How can the cost function be constructed from these input demand correspondences? For expository purposes and empirical illustration of the next section, consider a model having three inputs. A generalization to the n>3 input case is straightforward. Let w denote the price of the ith input,

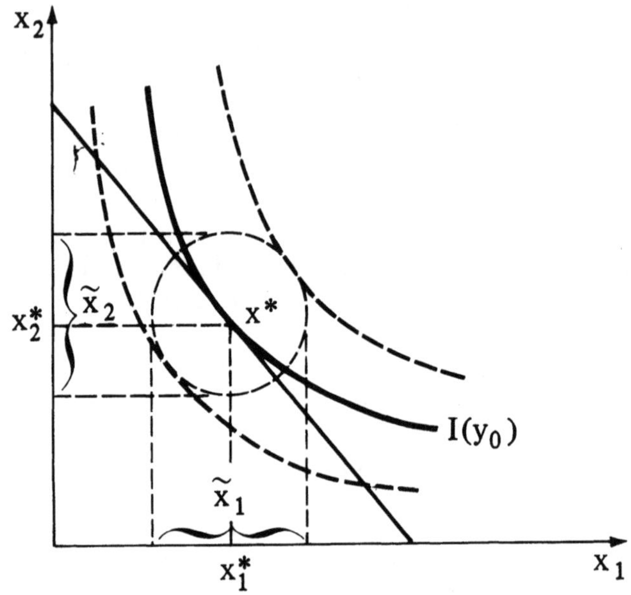

Figure 1: $\tilde{x}_1$ and $\tilde{x}_2$, a crisp isoquant $I(y_0)$ for the output level y_0, and a thick fuzzy isoquant.

[See equations (2.3) and (2.4)]

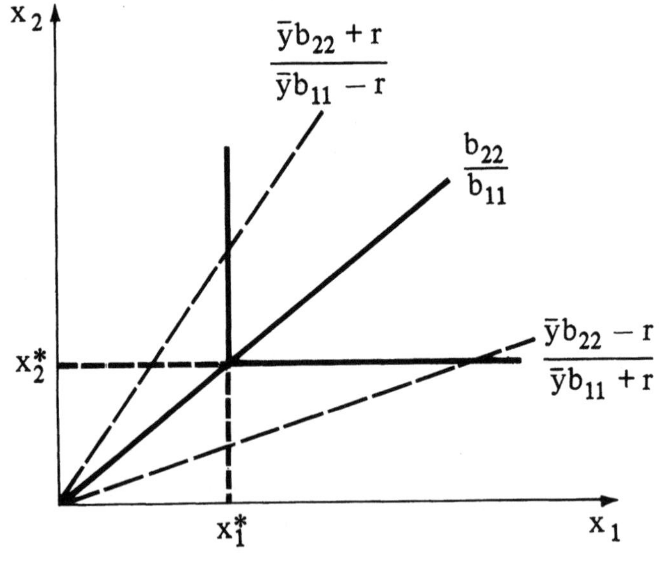

Figure 2: Crisp expansion path and its fuzzy neighborhood from equation (2.21).

where i=1,2,3. In a crisp technological environment the conditional cost function, $C^*(w_1, w_2, w_3, y)$ is obtained by solving the following minimization problem:

$$\text{Minimize } w_1x_1 + w_2x_2 + w_3x_3 \tag{2.9}$$

$$\text{subject to } f(x_1,x_2,x_3) = y \tag{2.10}$$

where $f(\cdots)$ is the crisp production function facing the firm. The cost function in this case basically represents the minimal cost for producing the given level of output, y. Using Shephard's Lemma, one obtains the conditional factor (input) demand functions as follows:

$$\partial C \;/\; \partial w_i = x^*(w_1, w_2, w_3, y) \tag{2.11}$$

This is the well known duality property of the cost function. The question is, how to modify this duality property in a fuzzy production environment? Let $\tilde{c}$ denote the total cost in the fuzzy technology. Since the input prices are observed in the market, it is reasonable to assume that these prices are crisp. We assume that the entire fuzz in $\tilde{c}$ arises due to a fuzz in input demands. Let us define,

$$\tilde{c}(w_1, w_2, w_3, y) = w_1\tilde{x}_1 + w_2\tilde{x}_2 + w_3\tilde{x}_3. \tag{2.12}$$

Using (2.6) to (2.8) our $\tilde{c}$ can be rewritten as:

$$\tilde{c} = C^* + \lambda_1 w_1 + \lambda_2 w_2 + \lambda_3 w_3 \tag{2.13}$$

where

$$\lambda_1 = r\cos\theta, \quad \lambda_2 = r\sin\theta\cos\phi, \quad \text{and} \quad \lambda_3 = r\sin\theta\sin\phi \tag{2.14}$$

By construction, λ_1, λ_2, and λ_3 in (2.14) must satisfy the following restriction.

$$\lambda_1^2 + \lambda_2^2 + \lambda_3^2 = r^2 \tag{2.15}$$

This is our econometric restriction which identifies the cost function, $\tilde{c}(\cdots)$ for a fuzzy technology. Thus, we suggest following steps. First consider a standard neoclassical cost function C^* and add the adjustment factor $\Sigma_{i=1}^{3}\lambda_i w_i$, which satisfies the identifying restrictions (2.15). Second, estimate our cost function for the fuzzy case in (2.13), subject to (2.15) by an appropriate estimator. Third, after estimating $\tilde{c}$ apply Shephard's Lemma to determine the fuzzy conditional factor demand functions by differentiating $\tilde{c}$ with respect to w_i. To illustrate the third step when n=3, use (2.6) to (2.8),

(2.12) and (2.14) to yield

$$\partial\tilde{c}/\partial w_i = \tilde{x}_i = x_i^* + \lambda_i \quad (2.16)$$

which immediately confirms (2.15) from the fact that

$$\Sigma_{i=1}^{3}(\tilde{x}-x_i^*)^2=1.$$

The fourth and final step is to apply duality mapping[2] to the fuzzy cost function (the same way we apply it to the crisp neoclassical cost and production functions) to recover the fuzzy technology using the duality mapping. We illustrate this step graphically in Figure 2 by considering the following Leontief cost function, with only two inputs (n=2):

$$C^*(w_1, w_2, y) = y\ [\, b_{11}w_1 + b_{22}w_2] \quad (2.17)$$

Clearly,

$$\partial C^*/\partial w_1 = x_1^* = b_{11}\, y \quad (2.18)$$

$$\partial C^*/\partial w_2 = x_2^* = b_{22}\, y \quad (2.19)$$

In the absence of a fuzz, the optimal input proportion stays on a ray through the origin with the slope b_{22}/b_{11}. With fuzz, the cost function is:

$$\tilde{c}(w_1, w_2, y) = y\ [\, b_{11}w_1 + b_{22}w_2] + w_1 r\cos\theta + w_2\, r\sin\theta \quad (2.20)$$

from (2.3) and (2.4). An application of the duality mapping yields:

$$\tilde{x}_2/\ \tilde{x}_1 = (y\ b_{22} + r\sin\theta)\ /\ (y\ b_{11} + r\cos\theta) \quad (2.21)$$

which defines the fuzzy neighborhood around the neoclassical efficient input frontier characterized by the ratio: b_{22}/b_{11}. Note that the input proportion (2.21) from a fuzzy cost function depends on the fuzz parameters θ, r, and the level of output y. Since $\tilde{x}_2/\ \tilde{x}_1$ now depends on the level of output, it is evident that the constant returns to scale property is violated when fuzz is introduced in the technology. This is not unexpected. In fact, Chen and Yu (1988) characterize many cases where constant returns to scale cease to hold. Also, the empirical results of Diewert and Wales (1987, p.60) provide pervasive rejection of constant returns to scale hypothesis.

Figure 2 depicts the crisp expansion path b_{22}/b_{11} and two extreme rays for input proportions (2.21) evaluated at $y=\bar{y}$, and $x_i^*=b_{ii}\ \bar{y}$, where $\bar{y}$ equals average output. Using $x_i^* - r \leq \tilde{x}_i \leq x_i^* + r$ stated above, it is clear that $\min(\tilde{x}_2/\ \tilde{x}_1) = (\bar{y}\ b_{22} - r)/(\bar{y}\ b_{11} + r)$ is the smallest value of the ratio, assuming that $b_{ii} \geq 0$. This is shown in Figure 2 as a ray on the right hand side of the crisp path for a small r. Similarly, $\max(\tilde{x}_2/\ \tilde{x}_1) = (\bar{y}\ b_{22} + r)/(\bar{y}\ b_{11} - r)$ is shown as a ray on the left hand side of the crisp expansion path. The shaded region between these two rays indicates the fuzz in the technology, which is why the firm cannot ex ante identify the crisp expansion path. Once we estimate the parameters θ and r, and know the output level y, we can know the ex post expansion path. In the next section we illustrate our estimation procedure using Nerlove's (1963) data and a generalized Leontief cost function having three inputs.

3. Estimation

This Section describes an empirical application of our model to Nerlove's(1963) famous cross sectional data set regarding Electric Utilities in 1955 in USA. The utility firms are divided into five groups of utilities with 29 observations each. The data have a total of 145 observations, and are explained in detail by Nerlove. They consist of $\tilde{c}_i$=Costs, y = Output, w_1= Wage Rate, w_2= Fuel Price, and w_3=Capital Price. Although our main interest is in the entire data set, we have included two subsets of 29 observations on larger utilities to assess the robustness of our methods and results, when applied to subsets. We have included a description of three sets of data: (i) the second largest 29 electric utilities from Nerlove's fourth group, (ii) the largest 29 utilities, and (iii) a set covering all 145 utilities from all five groups in our analysis. A flexible neoclassical cost function is specified to be generalized Leontief, defined by the following.

$$\tilde{c} = C^* + \lambda_1 w_1 + \lambda_2 w_2 + \lambda_3 w_3 \qquad (3.1)$$

where the crisp component of the cost $\tilde{c}$ in a fuzzy technological environment is

$$C^* = y\ [b_{11}w_1 + b_{12}\surd w_1 \surd w_2 + b_{13}\surd w_1 \surd w_3 + b_{22}w_2 + b_{23}\surd w_2 \surd w_3 + b_{33}w_3] \qquad (3.2)$$

Recently Diewert and Wales(1987) have used a similar function with additional terms for t, ty, y^2, t^2y, where t represents a trend, obviously absent in our cross sectional context. The term $\Sigma_{i=1}^{3}\lambda_i w_i$ has a new rationale based on the

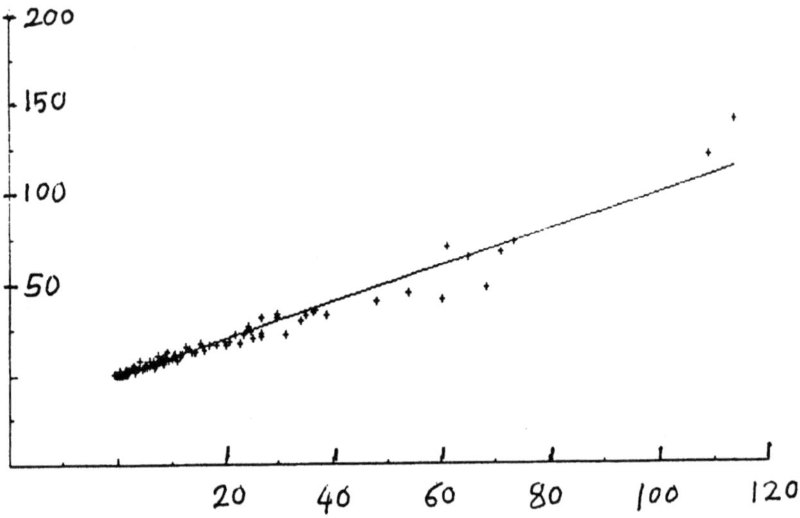

Figure 3: OLS fit with the observed cost on the horizontal axis and the fitted on the vertical.

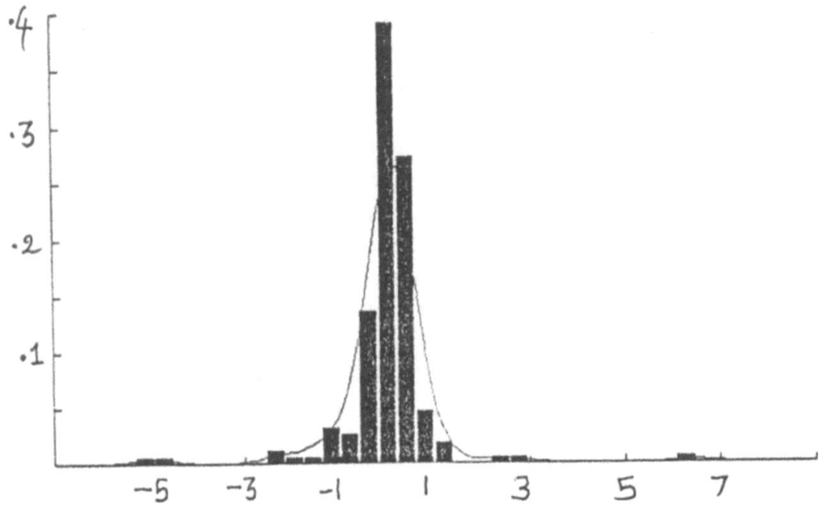

Figure 4: An empirical distribution of OLS residuals by Kernel methods and a superimposed histogram.

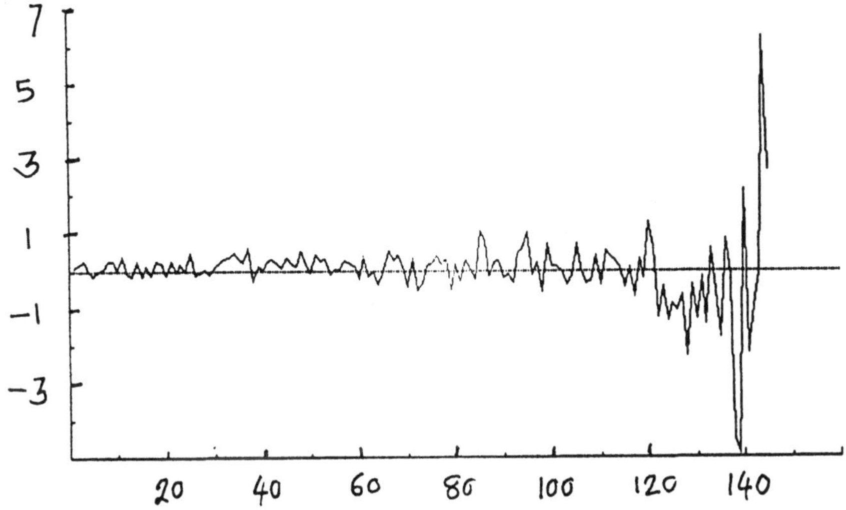

Figure 5: A plot of OLS residuals scaled by $\hat{\sigma}$ their sample standard deviation.

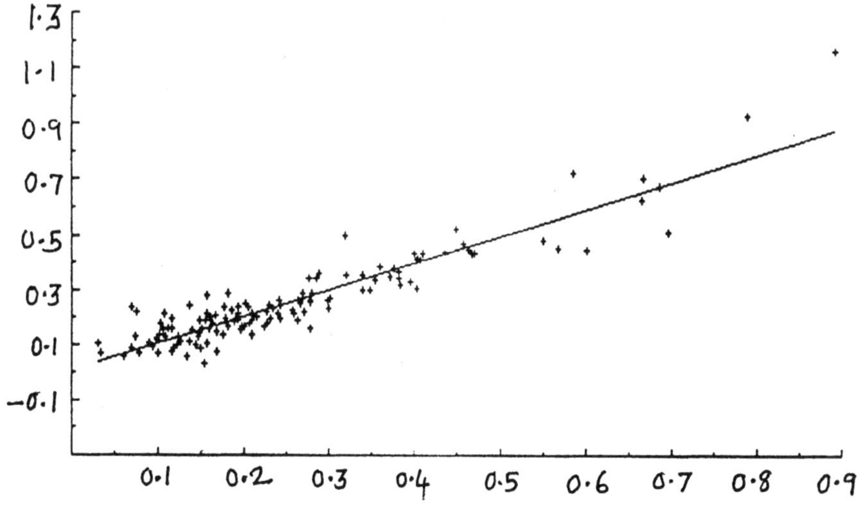

Figure 6: OLS fit with the observed cost on the horizontal axis and the fitted on the vertical for the model when all variables are divided by $\sqrt{y}$ to correct for heteroscedasticity.

fuzz in the input demand functions. Observe that this term makes (3.1) no longer separable in y, implying that the scale elasticity ϵ defined in (3.5) below need not be unity – obviously appropriate for electric utilities studied here. Intuitively, a scale elasticity ϵ=1 means that the output changes by one percent when all inputs change by one percent. A special case used here is

$$\tilde{c} = C^* + \lambda_1 w_1 + \lambda_2 w_2 + \lambda_3 w_3 \ , \quad \text{with } \lambda_1^2 + \lambda_2^2 + \lambda_3^2 = r^2 \quad (3.3)$$

Given Nerlove's data we estimate the coefficients in (3.1) to (3.3) by ordinary least squares (OLS) after appropriately adding an error term. A graph of the OLS fit with the observed values of the cost on the horizontal axis and the fitted values on the vertical axis is given in Figure 3. The R^2 for the fit is 0.9714, rather high for cross sectional data. Our initial study of the residuals from OLS regression alerted us to the possibility of heteroscedasticity in these cross sectional data. Figure 4 plots an empirical distribution of OLS residuals, where a histogram has been superimposed. Kernel estimation of density as in Silverman (1986) is used for smoothing the empirical density. Kernel estimation is a relatively new nonparametric technique which starts with the empirical histogram, and the kernel is a weighting function (usually the normal distribution) which integrates to unity. As explained in Silverman, one divides the range of available data into intervals of width $\sigma/4$ (say) if σ is the standard deviation of the data. For further applications of kernel methods in econometrics the reader is referred to Vinod (1988). Figure 4 shows a leptokurtic distribution with non-normal tails. The Jarque-Bera(1980) chi-square test (with degrees of freedom 2) for normality of errors yields a value of 1865.647, which is clearly significant supporting the visual impression of non-normality. White's(1980) F test for heteroscedasticity uses the R^2 from an auxiliary regression of squared residuals on all squares and cross products of original regressors. The null hypothesis is that errors are homoscedastic or, if heteroscedasticity is present it is unrelated to the regressors. Since F(18,116)=26.8179, the null hypothesis is rejected. Figure 5 also gives a plot of OLS residuals scaled by $\hat{\sigma}$ (full sample equation standard deviation of residuals), which further confirms increasing errors for larger electric utilities. The Goldfeld-Quandt F test for heteroscedasticity compares the residual sum of squares of the top and bottom set of observations after omitting some observations in the middle. Omitting 10 observations the observed

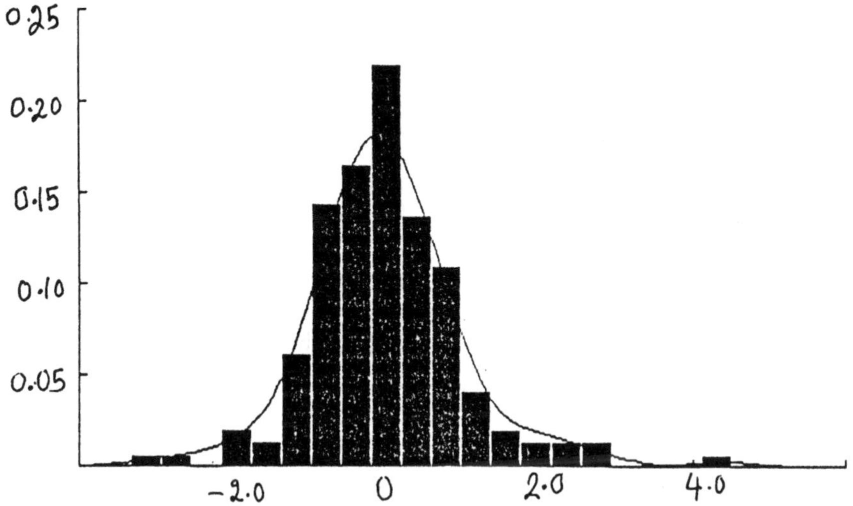

Figure 7: An empirical distribution of OLS residuals by Kernel methods and a superimposed histogram for the model when all variables are divided by $\sqrt{y}$ to correct for heteroscedasticity.

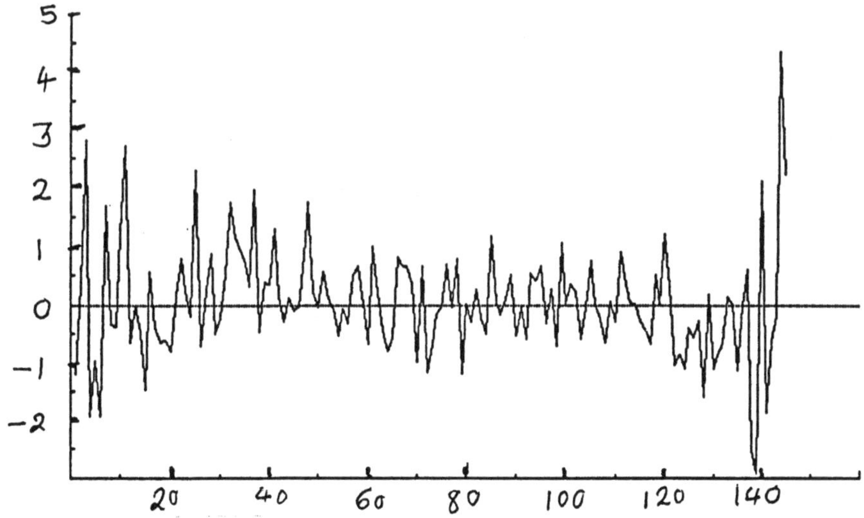

Figure 8: A plot of OLS residuals scaled by ($\hat{\sigma}$) their sample standard deviation for the model when all variables are divided by $\sqrt{y}$ to correct for heteroscedasticity.

F(57,60)=64.09 also rejects homoscedasticity.

How to solve the heteroscedasticity? Assuming that the variance is proportional to the output y, we transform all variables in (3.1) by dividing each of them by $\surd y$. This is the well known square root transformation to correct for heteroscedasticity. The OLS results after this transformation are reported in Tables 1 to 3 for the three data sets mentioned above. It is the results in Table 3 that refer to all 145 observations that should be compared to the results discussed in the previous paragraph. Figure 6 plots the line of regression similar to Figure 3 for the heteroscedasticity corrected model. Figure 7 plots the kernel estimation of residual density similar to Figure 4, suggesting distribution closer to normality. The Jarque-Bera $\chi^2(2)$=58.710 suggests a major reduction from the above value 1865.647. However, we have not achieved normality. Similarly, there is a major reduction in White's F statistic for testing heteroscedasticity to 26.8179, which suggests that some heteroscedasticity remains present. By contrast, Goldfeld-Quandt F((57,60)=1.0922 suggests that we have eliminated the heteroscedasticity problem. Figure 8 depicts the scaled residuals similar to those in Figure 5 confirming near elimination of heteroscedasticity.

Recall that the introduction of fuzz gives rise to the additional term, $\lambda_1 w_1 + \lambda_2 w_2 + \lambda_3 w_3$ in the cost function. The neoclassical cost function becomes a special case when λ_i=0 for i=1,2,3. Whether to use C^* or $\tilde{c}$ is an empirical question, which may be determined by an F test on the null hypothesis that λ_i=0 for i=1,2,3. Tables 1B, 2B and 3B report corresponding OLS estimates for crisp cost functions subject to the restriction that $\lambda_1=\lambda_2=\lambda_3=0$ along with the F statistic for this restriction and the tail areas. The F(3,136)=1165675.3 value is highly significant in Table 3B for the main data set of interest, based on all 145 observations. For the subsets in Tables 1B and 2B the F values are also large, F(3,20)=29014.156 and 31267.66, respectively. We conclude that the presence of fuzz is supported by the data.

In these tables, some individual regression coefficients are statistically insignificant. However, omitting them changes the underlying economic model specification, and affects all coefficients. We follow the approach of considering the plausibility of economic properties, of marginal costs, partials of cost with respect to prices, sign of the Hessian determinant, and so forth. We do not emphasize the signs and magnitudes of individual coefficients. There are some negative coefficients in all three tables, and one may wonder whether they are consistent with economic theory. The theory requires that the empirical estimates of partial derivatives of cost with respect to the input prices should be positive. As a practical matter these partials have to be evaluated at specific values of the right hand side variables in (3.1) to (3.3), which are usually chosen to be the sample means, denoted by bars. For example, we require that

$$\partial\tilde{c}/\partial w_1=\bar{y}[\ b_{11}+b_{12}(\surd\bar{w}_2/2\surd w_1)+b_{13}(\surd\bar{w}_3/2\surd\bar{w}_1)]+\lambda_1 \qquad (3.4)$$

where b_i are replaced by corresponding estimates from the three tables. Similarly, we evaluate the partials with respect to w_2 and w_3, and find to be positive in all cases. The partial derivative obtained by substituting the parameter estimates of Table 3 in equation (3.4) is 3.2426. Similarly, the partials with respect to w_2 and w_3 are 0.2807 and 0.0231 respectively.

The marginal cost(MC)=$\partial\tilde{c}/\partial y$, is simply the bracketed expression in (3.2). From the estimates in Table 3 for the fuzzy cost function we estimate MC=0.006. The average cost(AC) = $\tilde{c}/y$ is estimated by the sample average of this ratio to be 0.0129. Following Ferguson(1971, Ch. 8) let $\tau_i=(\partial x_i/\partial w_i)(w_i/x_i)$ denote the elasticity of input supply. Let $\bar{\tau}$ denote a weighted average of these τ_i values, with weights equal to the expenditure on the inputs. We cannot estimate $\bar{\tau}$, because the data on input quantities are not available to us. Economic theory suggests that these price elasticities τ_i should be negative, whence $\bar{\tau}$ is generally negative. Now, the scale elasticity ϵ is related to the MC, AC and $\bar{\tau}$ by the following relation proved by Ferguson(1971, p.179).

$$\epsilon = \frac{AC}{MC}\left(1 + \frac{1}{\bar{\tau}}\right) \qquad (3.5)$$

From the above estimates, AC/MC=2.1494, and the scale elasticity ϵ must be generally smaller than 2.1494 ($\bar{\tau}<0$). This is consistent with Nerlove's estimate that $\epsilon > 1$, based on a different model. Since electric utilities are likely to be so-called natural monopolies, it is not surprising that they have economies of scale. We have noted following (3.2) that ϵ need not be unity, and our cost function is not separable due to the presence of a fuzz in input demand functions.

Another question is whether the estimated cost function is concave, which depends on the second order partials. For example,

$$\partial^2 \tilde{c}/\partial w_1^2 = \bar{y}\ [\ \frac{-1}{4}\ b_{12}\ \sqrt{\bar{w}_2}\ \bar{w}_1^{-1.5} + \frac{-1}{4}\ b_{13}\ \sqrt{\bar{w}_3}\ \bar{w}_1^{-1.5}\ ] \quad (3.6)$$

The cross partials are appropriately defined. We report the Hessian matrix and its eigenvalues. The concavity of any twice differentiable function is assured if the determinant (product of eigenvalues) is non-negative. Table 3 reports the following eigenvalues: -2.4194, 0, and 0.0096. Thus we can expect the true Hessian determinant to be either zero, or negative. Hence, the estimates based on the fuzzy cost function of (3.3) are consistent with economic theory.

Detailed results are contained in Tables 1 to 3, including 1B to 3B. For brevity, let us only highlight the main results. A comparison of first pages of Tables 1, 2, and 3 with their neoclassical (crisp) counterparts in Tables 1B, 2B, and 3B reveal the following pattern. The standard error of estimate is lower for the fuzzy model than the crisp. The fit is generally superior for the fuzzy model. This can also be seen from the fact that R^2 values adjusted for degrees of freedom are always larger in Tables without the suffix B. The first-order partials of cost with respect to the three prices are always positive for the fuzzy model, whereas some wrong signs are observed for the crisp. We report the eigenvalues of the Hessian matrix of second order partials in all cases. They reveal a large negative eigenvalue in all cases where the fuzzy specification is used. The computations are based on the GAUSS computer package, which uses some 18 digits of accuracy. When an eigenvalue has many zeroes to the right hand side of the decimal point, it is zero for the numerical accuracy of the computer and for all practical purposes. For the fuzzy model we find that the eigenvalues are either close to zero or large and negative. Our overall results suggest that the cost function estimates continue to be consistent with economic theory. Furthermore, the behavior of eigenvalues is generally superior for the fuzzy than for the crisp model.

Our next problem is to further understand and interpret the fuzz parameters using the reduced form coefficients, λ_i. Since the fuzz factor is represented by a spheroid around the standard neoclassical profit maximizing equilibrium, x^*, our problem is to identify the parameters characterizing the precise form of the spheroid. It is important to observe that we do not need to impose any explicit restrictions on λ_1, λ_2, and λ_3 while estimating the flexible cost function. Thus, the estimation part is clearly separated from the identification of the fuzz. Using unrestricted estimates of λ_1, λ_2 and λ_3 we can estimate the fuzz parameters r, θ, and ϕ as follows.

$$r=\surd(\ \lambda_1^2 + \lambda_2^2 + \lambda_3^2\ ),\ \text{satisfying}\quad \lambda_1^2 + \lambda_2^2 + \lambda_3^2 = r^2 \quad (3.7)$$

and

$$\phi = \tan^{-1}(\ \frac{\lambda_3}{\lambda_2}), \qquad \text{and} \qquad \theta = \cos^{-1}(\frac{\lambda_1}{r}\ ) \qquad (3.8)$$

Tables 1 to 3 report the estimated radius r and angles (in radians) θ and ϕ based on (3.7) and (3.8). The sphere makes these angles with the corresponding axes. The angles in degrees, which are not reported here, are obtained by multiplying the radians by $180/\pi$. The estimates of λ_i in Table 3 satisfy $\lambda_1>\lambda_2>\lambda_3$, implying a larger fuzz in the cost of labor than in the cost of fuel, which in turn is somewhat larger than the fuzz in the cost of capital. These results are intuitively appealing, since the measurement of the labor input is likely to be least reliable.

Figure 9 depicts a fuzzy isoquant revealed by the data for 145 observations. The origin is shifted to the crisp input quantities x_1^*, x_2^*, x_3^*. Using the estimates of λ_1, λ_2, and λ_3 substituted in (3.7) and (3.8), we obtain the fuzz parameters as spherical coordinates. r= 0.1106, θ= 1.5949 in radians or 12.30°, and ϕ= 0.2147 in radians or 1.3628°. Table 3 reports spherical coordinates immediately following the OLS estimates. The point P in Figure 9 is located at the observed point with coordinates ($\tilde{x}_1$, $\tilde{x}_1$, $\tilde{x}_1$). The fuzzy isoquant is shown as a shaded solid (surface of revolution) traced by a sphere passing through P. The presence of fuzz is supported by the data and may be attributed to the inability of the firm to pay the inputs exactly according to the neoclassical optimization and duality theory.

4. Concluding Comments

This paper provides a new practical framework for estimating the cost function incorporating fuzziness in the technological environment. We construct spherical neighborhood around the usual profit maximizing equilibrium to allow for fuzzy isoquants. We demonstrate that this generalization causes very little alteration in the duality property between the cost function and the technology. The model proposed here is empirically estimable. The novelty of our method is in avoiding any complex nonlinear estimation in the reduced form parameters. Using certain unrestricted

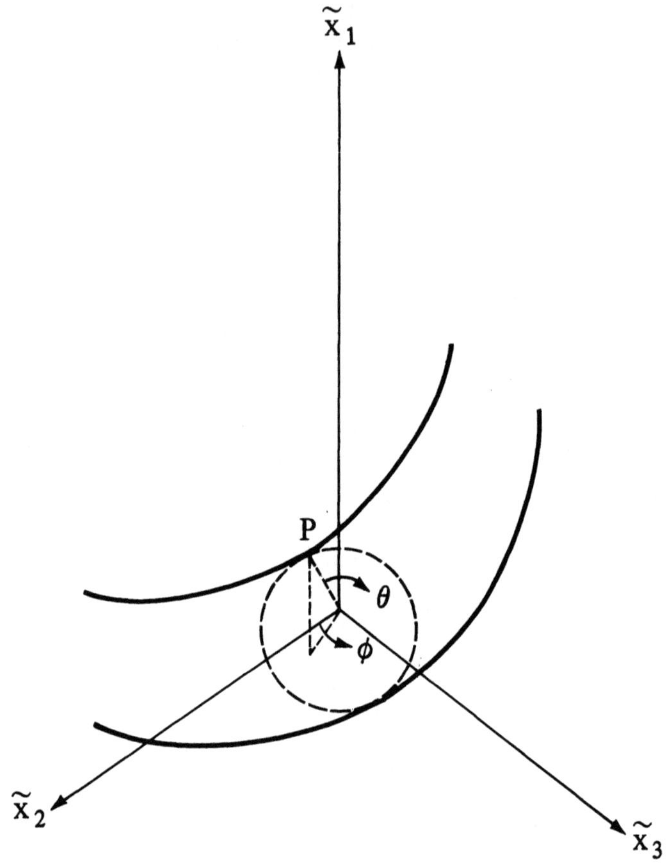

Figure 9: A fuzzy isoquant revealed by Nerlove's data (145 observations, Observed r = 0.1105, θ = 12.30°, and ϕ = 1.3628°)

coefficients, we can identify the fuzz by translating their values into the corresponding spherical coordinates. The estimation results are found to be consistent with the basic requirements of the neoclassical theory. Because marginal costs are positive, scale elasticity is greater than unity for public utilities which are likely to be natural monopolies. The cost function is concave in factor prices from the Hessian. A study of the entire sample and two subsets reveals that fuzzy estimation is superior to crisp. Our methodology can be easily extended to larger models and used for many other empirical models where some fuzz is expected.

APPENDIX

Use of Sines and Cosines for explicit expressions for Coordinates

The purpose of this appendix is to explain some basic relations involving sines and cosines used in the text and their generalizations for higher dimensions. Equation (2.5) and (2.14) of the text may be generalized to n dimensions by writing

$$\Sigma_{i=1}^{n} (\tilde{x}_i - x_i^*)^2 r_i^{-2} = 1 \qquad (A.1)$$

which defines an ellipsoid. The spherical spaces considered in the text are special cases with r_i=1 for all i. We wish to write $\tilde{x}_i = x_i^* \pm z_i$, where z_i is explicitly known. In $\Re^2$ if we choose $z_1 = r_1\cos\theta_1$ and $z_2 = r_2\sin\theta_1$, we can readily verify (A.1) by using the familiar identity with i=1.

$$\sin^2\theta_i + \cos^2\theta_i = 1 \qquad (A.2)$$

In $\Re^3$, we need θ_2 in addition to θ_1; and the three explicit expressions for z_i are as follows: $z_1 = r_1\cos\theta_1$, the same as for $\Re^2$, $z_2 = r_2\sin\theta_1\cos\theta_2$, and $z_3 = r_3\sin\theta_1\sin\theta_2$. To verify (A.1) in $\Re^3$, we use (A.2) for i=1,2. For the n=4 dimensional case, we have: $z_1 = r_1\cos\theta_1$, $z_2 = r_2\sin\theta_1\cos\theta_2$, $z_3 = r_3\sin\theta_1\sin\theta_2\cos\theta_3$, and $z_4 = r_4\sin\theta_1\sin\theta_2\sin\theta_3$. For n=5 (z_i/r_i) for i=1 to i=5 are respectively: C_1, S_1C_2, $S_1S_2C_3$, $S_1S_2S_3C_4$ and $\Pi_{j=1}^{n-1}S_j$, denoting $S_i = \sin\theta_i$ and $C_i = \cos\theta_i$ for brevity. Adding one more dimension after n-1 to n requires one more parameter θ_{n-1}, z_n then becomes $r_n\Pi_{j=1}^{n-1}S_j$, having one more term, and $z_{n-1} = r_{n-1}\Pi_{j=1}^{n-2} S_jC_{n-1}$. Since the number of fuzzy variables in a model may be large, it is useful to have the possibility of explicitly introducing them.

End Notes

1. For our purpose, we assume throughout that the input prices facing the firm are always crisp. This is justified on the grounds that the firm's choice of input levels is ex ante in nature, while the input prices are observed ex post in the market. In an ex ante sense, of course, the firm is vague regarding the efficient input choice, given the observed prices.

2. It is important to realize that the well-known techniques of Shephard's Lemma and duality mapping are applicable in the present context because the fuzz factor in equation (2.13) appears in an additive fashion. In general there may be complex interaction between the crisp and fuzz components of the cost function. More research is needed to examine how the duality mappings generalize to this situation.

REFERENCES

Blin J.M. et al. (1974), "Pattern Recognition In Microeconomics, Journal of Cybernetics, 3- 4, 17-27.

Chen, G. and E.S.H. Yu (1988), "On the Properties of Fuzzy Production Correspondence," Fuzzy Sets and Systems 27, 115-129.

Chang (1977), "Application of Fuzzy Set Theory to Economics, Cybernetics 6, 203-207.

Diewert, W.E. and T.J. Wales (1987), "Flexible Functional Forms and Global Curvature Conditions," Econometrica 55, 43-68.

Ferguson, C. E.(1971) The Neoclassical Theory of Production and Distribution London: Cambridge University Press.

Hirshleifer, J. and J.G. Riley (1979), "The Analytics of Uncertainty and Information- An Expository Survey, Journal of Economic Literature, 17, 1375-1421.

-------- (1981), " The Analytics of Uncertainty and Information", Ch. 1, UCLA working Paper.

Jarque, C.M. and A.K. Bera (1980),"Efficient Tests for Normality, Homoscedasticity, and Serial Independence of Regression Residuals," Economics Letters 6, 255-259.

Kotz and Johnson (1988) Encyclopedia of Statistical Sciences, Wiley, NY.

Lester, R.A. (1947), "Marginalism, Minimum Wages and Labor Markets," American Economic Review 37, 148-154.

McCain, R. (1987) "Fuzzy Confidence Intervals in a Theory of Economic Rationality," Fuzzy Sets and Systems 23, 205-218.
Nerlove, M. (1963) "Returns to Scale in Electricity Supply," in Measurement in Economics Carl F. Christ (ed.) Stanford, California: Stanford University Press.
Ponsard, C (1982), "Producer's Spatial Equilibrium with Fuzzy Constraints," Journal of Regional Science 22, 159-175.
Popper, K (1973), "Of Clouds and Clocks," in Objective Knowledge: an Evolutionary Approach, Oxford: Oxford University Press, 206-55.
Silverman, B.W. (1986) Density Estimation for Statistics and Data Analysis London: Chapman and Hall.
Tanaka, H., S. Vejima and K. Assai (1982), "Linear Regression Analysis with Fuzzy Model," IEEE Transa ctions on Systems, Man, and Cybernetics 12, (6), 903-907.
Taranu, C (1977), "The Economic Efficiency- A Fuzzy Concept," Modern Trends in Cybernetics and Systems, Vol.2, Springer Verlag, New York, 163-173.
Vinod, H. D.(1988) "Random Walk in Consumption: Maximum Likelihood and Nonpara-metrics," in T.B. Fomby and G.F. Rhodes, Jr. (ed.s) Advances in Econometrics, Robust and Nonparametric Statistical Inference, Vol. 7, Greenwich, Conn.: JAI Press, 291-309.
Vinod, H.D. and P. Basu (1989) "Rational Expectations Equilibrium With Fuzzy Sets and Estimation," presented at Association of Indian Economic Studies Meeting in June at Montclair State College, New Jersey.
White, H. (1980), "A Heteroscedasticity Consistent Covariance Matrix Estimator and a Direct Test for Heteroscedasticity," Econometrica 48, 817-838.
Zadeh, L.A. (1965), "Fuzzy Sets," Information and Control 8, 338-353.

TABLE 1 . Nerlove's data on TYPE 4 Utilities.
Generalized Leontief Cost Function of equation (3.1) estimated after all variables are divided by $\surd y$ for heteroscedasticity.
Valid cases: 29 Dependent variable: cost/$\surd$y
Total SS: 0.139 Degrees of freedom: 20
Residual SS: 0.018 Std error of est: 0.030
F(9,20): 246.369 Probability of F: 0.000
Durbin-Watson (approximate indication of specification): 2.385
R^2 adjusted for degrees of freedom = 0.8122

Variable	Estimate	Standard Error	t-value	Prob >|t|	Std.zed Estimate
yw11	-0.015964	0.013010	-1.227056	0.234	-3.715331
yw12	0.000488	0.002935	0.166260	0.870	0.608632
yw13	0.004513	0.002253	2.002692	0.059	7.916170
yw22	-0.000084	0.000225	-0.372312	0.714	-0.518483
yw23	-0.000003	0.000253	-0.011736	0.991	-0.031140
yw33	-0.000276	0.000124	-2.223026	0.038	-4.480071
w1	-12.02573	6.025707	-1.995737	0.060	-1.086989
w2	0.309235	0.209913	1.473159	0.156	0.841426
w3	0.104669	0.055132	1.898509	0.072	0.698730

Variable names mean the following: yw11=$w_1\surd y$, yw12= $\surd(yw_1w_2)$, yw13= $\surd(yw_1w_3)$, yw22= $w_2\surd y$, yw23=$\surd(yw_2w_3)$, yw33= $w_3\surd y$, w1=$w_1/\surd y$, w2= $w_2/\surd y$, and w3=$w_3/\surd y$.
The last three coefficients estimate λ_1, λ_2, and λ_3 respectively.
Spherical coordinates= 12.030159 1.5620957 -0.025708807
First partials with respect to w_1 to w_3: 1.4129033 0.28328179 0.22830815
MC= 0.0049396499 AC= 0.0054513114
ϵ = Scale elasticity > average cost/marginal cost = 1.1035826
Hessian matrix of second order partials=

-12.639071	0.043063355	0.13799963
0.043063355	-0.0038298834	-2.7851115E-005
0.13799963	-2.7851115E-005	-0.0015598713

determinant= 0
eigenvalues of Hessian
-12.640725, -0.0037362666, and 0.

TABLE 1B . Nerlove's data on TYPE 4 Utilities.

Generalized Leontief (Crisp) Cost Function subject to
$\lambda_1 = \lambda_2 = \lambda_3 = 0$

Valid cases:	29	Dependent variable:	$cost/\surd y$
Total SS:	459.433	Degrees of freedom:	23
Residual SS:	79.247	Std error of est:	1.856
F(6,23):	224.891	Probability of F:	0.000

Durbin-Watson (approximate indication of specification): 1.335
R^2 adjusted for degrees of freedom = 0.7782

Variable	Estimate	Standard Error	t-value	Prob >|t|	Std.zed Estimate
yw11	-0.560107	0.791919	-0.707278	0.486	-2.264826
yw12	0.036174	0.179439	0.201596	0.842	0.783775
yw13	0.118720	0.130878	0.907106	0.374	3.618035
yw22	-0.000033	0.012668	-0.002592	0.998	-0.003528
yw23	0.000150	0.015504	0.009667	0.992	0.027343
yw33	-0.006097	0.006698	-0.910327	0.372	-1.721075

Variable names mean the following: $yw11 = w_1\surd y$, $yw12 = \surd(yw_1w_2)$, $yw13 = \surd(yw_1w_3)$, $yw22 = w_2\surd y$, $yw23 = \surd(yw_2w_3)$, and $yw33 = w_3\surd y$.

First partials with respect to w_1 to w_3:
124.45598 13.039584 0.53049951
MC= 0.25752424 AC= 0.0054513114
ϵ= Scale elasticity > average cost/marginal cost = 0.021168148
Hessian matrix of second order partials=

-354.31599	3.1918684	3.6302461
3.1918684	-0.31278143	0.0014075903
3.6302461	0.0014075903	-0.041291326

determinant= 0
eigenvalues of Hessian
-354.38195 , -0.28810678, and 0.
Unrestricted URSS= 0.018204667 Restricted RRSS= 79.247160
F(3,20) statistic= 29014.156 for testing $\lambda_1 = \lambda_2 = \lambda_3 = 0$

TABLE 2 . Nerlove's data on TYPE 5 Utilities,

Generalized Leontief Cost Function
of equation (3.1) estimated after all variables are divided by $\sqrt{y}$
for heteroscedasticity.
Valid cases: 29 Dependent variable: cost/$\sqrt{y}$
Total SS: 1.167 Degrees of freedom: 20
Residual SS: 0.164 Std error of est: 0.090
F(9,20): 103.014 Probability of F: 0.000
Durbin-Watson (approximate indication of specification): 1.639
R^2 adjusted for degrees of freedom = 0.7962

Variable	Estimate	Standard Error	t-value	Prob >|t|	Std.zed Estimate
yw11	-0.018771	0.016422	-1.143032	0.267	-4.633971
yw12	-0.000647	0.005642	-0.114631	0.910	-0.581994
yw13	0.004866	0.003013	1.614758	0.122	9.923081
yw22	-0.000261	0.000642	-0.406114	0.689	-0.913082
yw23	0.000153	0.000501	0.305314	0.763	1.193067
yw33	-0.000257	0.000191	-1.347311	0.193	-4.746024
w1	-10.45249	11.19872	-0.933363	0.362	-0.293637
w2	1.765067	0.877003	2.012611	0.058	0.893050
w3	-0.186092	0.142510	-1.305815	0.206	-0.460471

Variable names mean the following: yw11=$w_1\sqrt{y}$, yw12= $\sqrt{(yw_1w_2)}$, yw13= $\sqrt{(yw_1w_3)}$, yw22= $w_2\sqrt{y}$, yw23=$\sqrt{(yw_2w_3)}$, yw33= $w_3\sqrt{y}$, w1= $w_1/\sqrt{y}$, w2= $w_2/\sqrt{y}$,
and w3= $w_3/\sqrt{y}$.
The last three coefficients estimate λ_1, λ_2, and λ_3 respectively.
Spherical coordinates= 10.602085 1.5883496 -0.16728781
First partials with respect to w_1 to w_3 : 8.4020905
0.71790496 1.9971354
MC= 0.0071829158 AC= 0.0057892669
ϵ = Scale elasticity > average cost/marginal cost , ϵ =
0.80597728
Hessian matrix of second order partials=

-36.661589	-0.15456435	0.44628017
-0.15456435	-0.014525318	0.0039251720
0.44628017	0.0039251720	-0.0057335947

determinant=0
eigenvalues of Hessian
-36.667674, -0.014174220 , and 0.

TABLE 2B . Nerlove's data on TYPE 4 Utilities,
Generalized Leontief (Crisp) Cost Function subject to
$\lambda_1=\lambda_2=\lambda_3=0$

Valid cases:	29	Dependent variable:	cost/$\sqrt{y}$
Total SS:	24031	Degrees of freedom:	23
Residual SS:	7644	Std error of est:	18.231
F(6,23):	33.354	Probability of F:	0.000

Durbin-Watson (approximate indication of specification): 0.899
R^2 adjusted for degrees of freedom = 0.5989

Variable	Estimate	Standard Error	t-value	Prob >|t|	Std.zed Estimate
yw11	2.261890	3.133705	0.721794	0.478	3.891794
yw12	-0.842575	1.115970	-0.755015	0.458	-5.284372
yw13	-0.096092	0.563806	-0.170435	0.866	-1.365883
yw22	-0.042992	0.119887	-0.358606	0.723	-1.049886
yw23	0.129643	0.098287	1.319023	0.200	7.054396
yw33	-0.021273	0.036530	-0.582337	0.566	-2.738735

Variable names mean the following: yw11=$w_1\sqrt{y}$, yw12= $\sqrt{(yw_1w_2)}$, yw13= $\sqrt{(yw_1w_3)}$, yw22= $w_2\sqrt{y}$, yw23=$\sqrt{(yw_2w_3)}$, and yw33= $w_3\sqrt{y}$.

First partials with respect to w_1 to w_3: 2190.4791
57.026304 -11.610462
MC= 0.54972925 AC= 0.0057892669
ϵ = Scale elasticity > average cost/marginal cost = 0.010531124
Hessian matrix of second order partials=

3330.6295	-201.35999	-8.8138036
-201.35999	-6.8150268	3.3299854
-8.8138036	3.3299854	-0.38870798

determinant=0
eigenvalues of Hessian
-19.332817 1.1723955E-013 3342.7586
Unrestricted URSS= 0.16371790
Restricted RRSS= 7644.1768
F(3,20) statistic= 311267.66 for testing $\lambda_1=\lambda_2=\lambda_3=0$

TABLE 3 . Nerlove's data on ALL TYPES of Utilities,

Generalized Leontief Cost Function of Equation (3.1), estimated after all variables have been divided by $\surd y$ to remove heteroscedasticity.

Valid cases: 145 Dependent variablecost/$\surd$y
Total SS: 3.807 Degrees of freedom136
Residual SS:0.540 Std error of est: 0.063
F(9,136): 346.132 Probability of F: 0.000
Durbin-Watson (approximate indication of specification): 1.813
R^2 adjusted for degrees of freedom = 0.8488

Variable	Estimate	Standard Error	t-value	Prob >|t|	Std.zed Estimate
$w_1\surd y$	-0.003029	0.008102	-0.373774	0.709	-1.079774
$\surd(yw_1w_2)$	-0.001646	0.002367	-0.695576	0.488	-2.087842
$\surd(yw_1w_3)$	0.001586	0.001389	1.141608	0.256	5.111382
$w_2\surd y$	0.000201	0.000205	0.976417	0.331	0.935758
$\surd(yw_2w_3)$	0.000114	0.000213	0.534390	0.594	1.312067
$w_3\surd y$	-0.000106	0.000082	-1.293518	0.198	-3.171123
$w_1/\surd y$	0.107984	0.180843	0.597114	0.551	0.147016
$w_2/\surd y$	0.023551	0.005143	4.579118	0.000	0.471532
$w_3/\surd y$	-0.002569	0.001877	-1.368840	0.173	-0.315470

The last three coefficients estimate λ_1, λ_2, and λ_3 respectively.
Spherical coordinates= 0.11055246 1.5940399
0.21473552
First partials with respect to w_1 to w_3 3.2426055
0.28065278 0.023101587
MC= 0.0059898574 , AC = 0.012874849
ϵ = Scale elasticity > average cost/marginal cost = 2.1494416
Hessian matrix of second order partials=

-2.4123784	-0.12220435	0.045595544
-0.12220435	0.0032299289	0.00089656222
0.045595544	0.00089656222	-0.00064979201

Determinant of the Hessian= 3.8539999E-020
Eigenvalues of the Hessian =-2.4194040 3.4694470E-018
0.0096056828

TABLE 3B. Nerlove's data on ALL TYPES of Utilities,

Generalized Leontief (Crisp) Cost Function subject to
$\lambda_1=\lambda_2=\lambda_3=0$

Valid cases: 145 Dependent variablecost/$\sqrt{}$y
Total SS: 56422 Degrees of freedom139
Residual SS: 13882 Std error of est: 9.994
F(6,139): 111.732 Probability of F: 0.000
Durbin-Watson (approximate indication of specification): 0.686
R^2 adjusted for degrees of freedom = 0.7433

Variable	Estimate	Standard Error	t-value	Prob >|t|	Std.zed Estimate
$w_1\sqrt{y}$	1.219684	1.282615	0.950935	0.343	3.571990
$\sqrt{(yw_1w_2)}$	-0.216152	0.374061	-0.577851	0.564	-2.251427
$\sqrt{(yw_1w_3)}$	-0.113667	0.220074	-0.516495	0.606	-3.008866
$w_2\sqrt{y}$	-0.083308	0.032330	-2.576771	0.011	-3.191811
$\sqrt{(yw_2w_3)}$	0.086938	0.033678	2.581448	0.011	8.245990
$w_3\sqrt{y}$	-0.011289	0.013017	-0.867301	0.387	-2.768449

First partials with respect to w_1 to w_3: 630.16091, 0.53891781, -1.4689461
MC= 0.41457532 , AC= 0.012874849
ϵ = Scale elasticity > average cost/marginal cost = 0.031055512
Hessian matrix of second order partials=

502.08120	-16.043126	-3.2676006
-16.043126	-3.3641702	0.68597645
-3.2676006	0.68597645	-0.065975884

Determinant of the Hessian=-2.9591943E-017
Eigenvalues of the Hessian
-3.9603549 1.3322676E-014 502.61140
Unrestricted URSS= 0.53987490, Restricted RRSS= 13882.573
F(3,136) statistic= 1165675.3, for testing $\lambda_1=\lambda_2=\lambda_3=0$,
the restriction imposed by crisp cost function .

ESTIMATING AND EXPLAINING THE COST OF HIGH-TECHNOLOGY SYSTEMS

0. Douglas Moses
Naval Postgraduate School

ABSTRACT

The purpose of this study is twofold: to demonstrate the use of technology measurement methodology in developing cost estimating relationships for high-technology systems and to identify environmental factors that impact the production cost of those systems. Both judgmental weighing and time regressions approaches are used to create measures of the state-of-the-art of technology and advances in technology at the time of system production. Those measures then are used to estimate system cost. Additionally, several political, economic and program factors are hypothesized to explain differences between actual and estimated costs. Tests of the hypotheses are conducted. Findings indicate that technology measurement provides a useful tool in developing cost estimates for high-technology systems and that environmental factors have a predictable impact on the actual cost of these systems. The analysis is conducted using a sample of U.S. military aircraft.

INTRODUCTION

Researchers and analysts are frequently concerned with the interrelationships between the "amount" of technology or the "advance" in technology in a system and the cost of producing that system. Alexander and Mitchell [1] have in fact noted the "striking similarity" between approaches for measuring technological change and cost-estimating relationships: both typically start with a set of variables representing multiple properties of systems and combine the variables in some manner to create measures of technology or measures of cost. Recent research [10, 13] has indicated that cost estimating relationships

---

* This research was sponsored by The Cost Estimating and Analysis Division, Naval Sea Systems Command and the Naval Postgraduate School.

for high-technology systems can be improved if, rather than estimating cost directly from system properties, the system properties are first "filtered" through a technology measurement procedure to create summary measures of system technology. Then the technology measures are used to estimate cost. This suggests that the various methodologies that have been developed to measure the state-of-the-art of technology and advances in technology [1,2,5,7,10,20] may serve a useful intermediate step in developing predictions of cost for new systems that extend the state-of-the-art. The first objective of this paper is to demonstrate the use of summary measures of the state-of-the-art of technology and advance in technology in developing cost estimating relationships. Two technology measurement approaches, a judgmental weighting [7] and a time regression [2,6,13], are used in this regard.

Actual costs of high-technology systems, however, are unlikely to be solely a function of the technological sophistication of the systems. Environmental factors--economic and political--are likely to impact the conditions in which new systems are developed and produced. If such factors are systematic and recurring, they can be expected to provide indications of when the cost of extending technology may be more or less than "normal". The second objective of this paper is to identify and test for the impact of various environmental factors on the cost of high-technology systems. Identifying such factors provides a first step in the possible incorporation of measures of these factors in future research attempting to predict the cost of extending technology.

The analysis was conducted using data for a sample of U.S. military aircraft produced during the 1950-1980 period. Since the choice of the approach for measuring technology in the systems relied in part on the data available for the sample, the next section describes sample selection procedures.

SAMPLE

The population for this study was originally defined as U.S. military aircraft. The sample represents a subset of military aircraft for reasons set out below. The source of data was the U.S. Military Aircraft Cost Handbook [3] produced under contract to the Department of Defense, containing performance and cost data on military aircraft manufactured from the early 1950's through the early 1980's.

The handbook contains data for 108 distinct individual aircraft, identified by mission (fighter, attack, patrol, bomber etc.), design and series. For example, the B-52C is a bomber (B), design (52), third series (C). Where successive series of a particular design resulted in virtually indistinguishable aircraft, the handbook combines series into a single program (e.g., A-7A, A-7B -->A-7A/B). This reduced the number of distinct aircraft programs to 80. Since the study is concerned with the state of technology represented by high-technology systems, as reflected in performance and capability (to be discussed in a later section), it was necessary to reduce the sample further. The methodology for assigning a performance measure to aircraft relies on a baseline aircraft, the F-4B, which is used in both fighter and attack missions. Aircraft designed for other missions (strategic bombers and patrol) were deleted (n=19). In addition, because the baseline F-4B is a conventional take-off-and-landing (CTOL) aircraft, and performance is related to the take-off-and-landing mode, vertical and short take-off-and-landing aircraft were deleted (n=6). Finally when successive series of a particular design had the same performance, it was assumed that no extension in technology had been achieved and the later series was deleted (n=8). Thus the final sample consists of 47 distinct CTOL fighter and attack aircraft manufactured from the early 1950's through the early 1980's. The list of the aircraft programs is available from the author.

MEASURING THE STATE-OF-THE-ART OF TECHNOLOGY

The literature on technology measurement offers various broad approaches to determining the state-of-the-art (SOA) of technology for a given set of related systems [1,2,5,7,9,10,20]. Each approach requires the knowledge of a number(n) of technology variables reflecting distinct properties or characteristics. Each approach combines the variables into a single SOA measure which has a scale independent of the scales of the individual technology characteristics (which are typically measured in differing types of units). The judgmental weighting approach expresses SOA as a direct combination of values of the technology characteristics. Gordon and Munson [9] suggest two general forms of SOA equations.

$$SOA = B_1V_1 + B_2V_2 + ... B_nV_n \quad \text{and}$$

$$SOA = V_1[B_2V_2 + B_3V_3 + ... B_nV_n]$$

where

B_i = judgmentally assigned weights

V_i = the value of the ith technology describing variable.

The first version of the model is a simple linear combination of weighted characteristics, the second version is a multiplicative form intended for use when one variable (V_1) must be present in the system.

The measures used to reflect technology in this study were constructed by The Analytic Sciences Corporation [23] and rely on the judgmental weighting approach. TASC determined two "figures of merit" for each aircraft. The airframe performance (AP) score reflects the performance and capability of the airframe and engine. The aircraft system performance (ASP) score reflects the capability of the airframe, engine and the electronics, navigation and weapons systems, i.e., the complete aircraft. Each score is a judgmentally weighted function of more basic properties.

Airframe performance is measured by

$$AP = B_1 \times P + B_2 \times R + B_3 \times M + B_4 \times V$$

where

B_1 = Judgmental weights
P = Payload
R = Range and basing mode
M = Maneuverability
V = Useful speed

This formulation is an additive multi-attribute utility function [15]. Because values of P, R, M, and V are expressed in different units, values for P, R, M and V for individual aircraft were divided by the corresponding values for the baseline F-4B aircraft. This results in all characteristics being expressed as ratios, which can be combined into an overall score. Weights were determined by the consensus judgment of a large panel of expert operational personnel. Weights were assigned such that the baseline F-4B had an AP score of 10.

Aircraft system performance is measured by

$$ASP = S\,(B_1 \times P \times U + B_2 \times R \times N + B_3 \times M + B_4 \times V)$$

where

S = Survivability modifier, reflecting susceptibility to detection, identification and destruction.

U = Payload utility modifier, reflecting target acquisition and target engagement capability.

N = Navigation coefficient, reflecting internal navigation system capability.

B_i,P,R,M,V = as previously defined.

Again, values of individual characteristics were scaled by the value for the baseline F-4B aircraft, and expert judgment was relied on for determining the functional form and weights of the utility function.

Note that the individual properties reflected in the models represents "output" measures of performance or capability along distinct dimensions. This is consistent with the work of Knight [16] who distinguishes between structural and functional technology measures. Structural measures capture physical characteristics, i.e., "what the system looks like". Functional measures capture capabilities, i.e., "what the system does". Measures of function or output can be used to compare systems of differing structure.

These two measures were taken as summary indicators of the SOA of technology embodied in the aircraft, reflecting their functional capability. Three technology SOA measures to be used in later analysis were defined as follows:

1. Platform (Airframe and Engine)Technology (PLATTECH) = AP.
2. Flyaway Aircraft System Technology (FLYTECH) = ASP
3. Weapons and Avionics System Technology (SYSTECH) = ASP/AP.

The SYSTECH measure is derived from the two others and is a rough attempt to capture the degree to which the technology in weapons systems and avionics systems enhances airframe and engine capability to achieve flyaway aircraft system capability. Dividing ASP by AP is consistent with the idea that the components in the AP formula have been multiplied by modifiers to arrive at ASP.

When speaking of the three technology measures collectively, the expression "TECH" will be used. Values of the TECH measures, and COST measures described later, are available from the author.

MEASURING EXTENSIONS OF TECHNOLOGY

Various researchers have developed methods for measuring extensions in technology (see Dodson [7] and Greer [10] for reviews). One common approach relies on the idea of the "year-of-technology" [2,13]. In this approach, time is related to technology measures in a multiple regression:

$$Y = a + b_1X_1 + b_2X_2 + \ldots + b_nX_n + e$$

where

Y = actual year the system become operational.
b_i = regression coefficients
X_i = technology measures
e = residual

A predicted value from the regression equation for an individual system represents the "year-of-technology" for that system. If the actual year a given system was produced is less than its year-of-technology, it can be said that the system was produced "ahead of its time" and represents an advancement in technology.

As simplistic as this method seems, related work by Lienhard [19] tends to support the concept. His paper studied the rate at which technology is improved, and how (whether) this rate changes through time. He studied several forms of technology (clocks, steam power, land transportation, low temperatures, air transportation) over extended time periods. The most relevant observation to come from Lienhard's study was that the rate of improvement of a particular technology, once established, does not change. If this is correct, there could be some major implications for the cost, and even the feasibility, of attempting to effect technological advances "before their time". If a desired advance could normally be expected to occur only by some quasi-naturally established date, attempts to accelerate this process would be very costly. Accordingly, the year-of-technology approach may be well reasoned.

The essence of the year-of-technology approach is to relate

technology to time and use deviations from the time line as indicators of the technology advancement represented by individual systems. A similar approach is used here, but TECH is treated as the dependent variable rather than time. (Since summary technology variables are used, rather than many technology characteristics, they can be used as the dependent variable with results that are equivalent but easier to display and discuss.)

Results from separately regressing the three TECH variables against the year in which the aircraft were first operational (YEAR) are shown in Table 1. Not surprisingly, in each case coefficients for YEAR are positive and significant, indicating that technology increases with time. The relatively high R^2 values indicate that time explains a large proportion of the technology variance among the aircraft.

TABLE 1
REGRESSION OF TECH ON YEAR

Dependent Variable	PLATTECH	SYSTECH	FLYTECH
Independent Variable	YEAR	YEAR	YEAR
Intercept	-8.619	-1.038	-30.943
YEAR Coefficient	.2971	.0362	.7063
Coefficient t	7.011	7.902	8.386
Significance	.0001	.0001	.0001
Model F	49.16	62.44	70.32
Model Significance	.0001	.0001	.0001
R2	.5221	.5812	.6098
Adjusted R^2	.5115	.5719	.6011

Plots of the three TECH measures over time are displayed in Figures 1, 2, and 3. The A-6E aircraft is highlighted in the figures. Observing Figure 1, we can see that the A-6E falls approximately on the trend line. The technology embodied in the A-6E platform was not in excess of the average state-of-the-art of platform technology at the time of the A-6E's production. Figure 2 shows that the weapons and avionic systems in the A-6E were advanced relative to the average state of systems technology. The result (Figure 3) was a flyaway aircraft also advanced relative to the average trend in aircraft technology at the time of the A-6E's production.

These observations can be generalized to define three variables

FIGURE 1
PLOT OF PLATFORM TECHNOLOGY OVER TIME

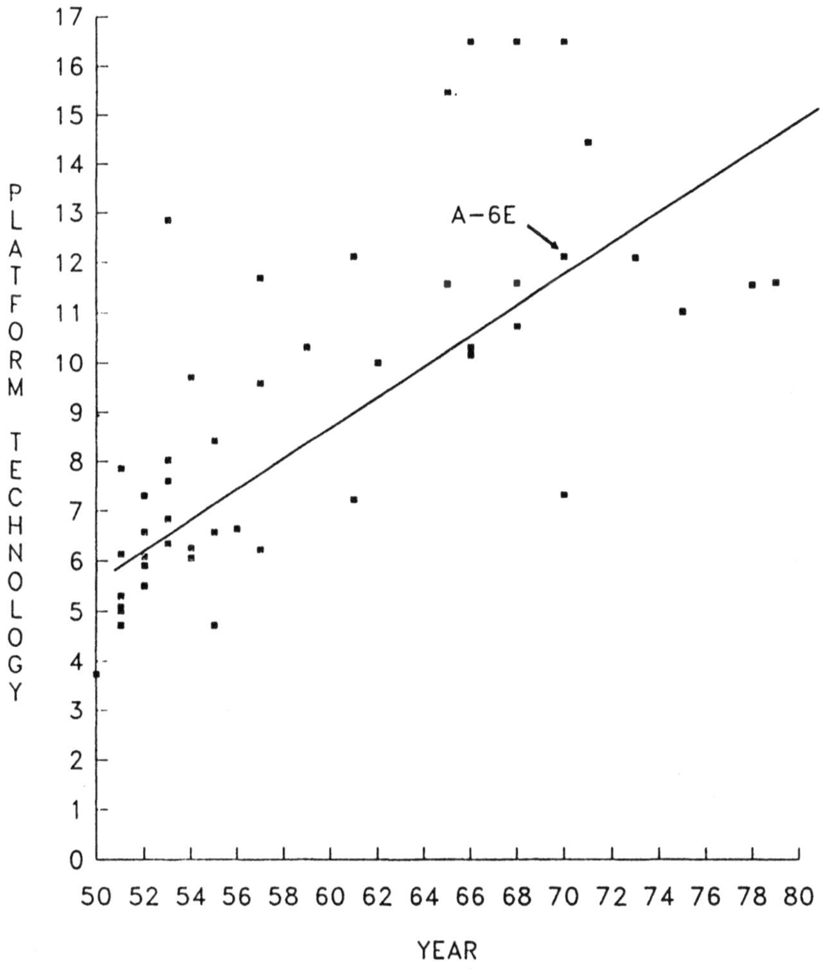

FIGURE 2
PLOT OF SYSTEM TECHNOLOGY OVER TIME

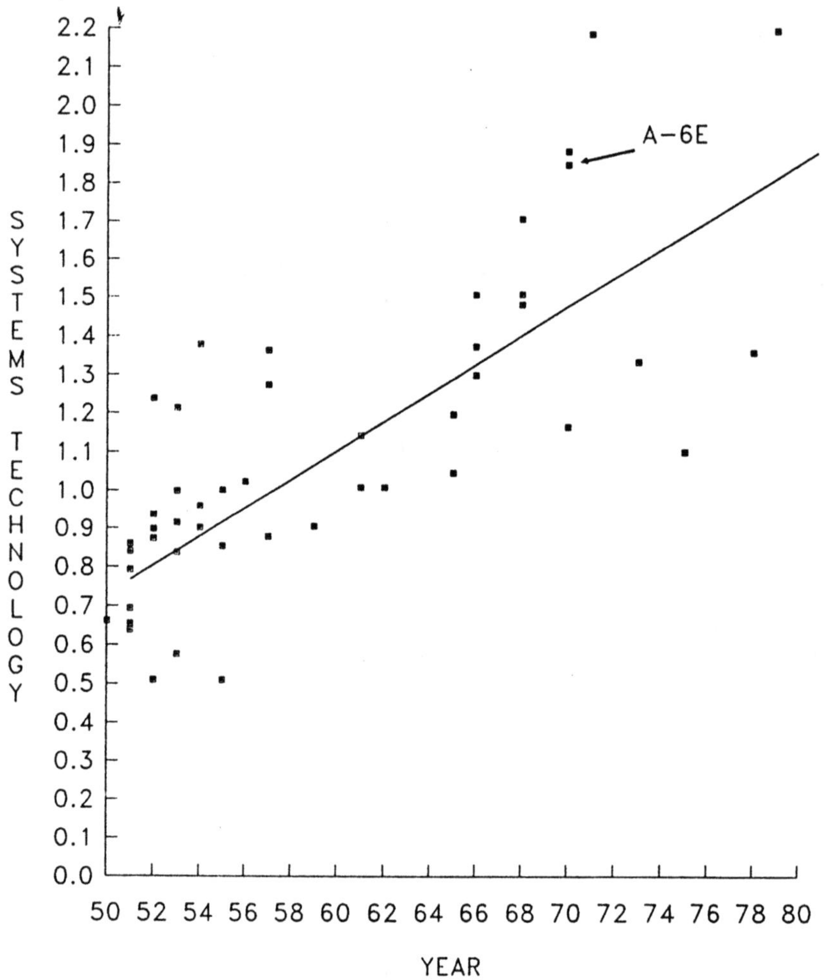

FIGURE 3
PLOT OF FLYAWAY AIRCRAFT OVER TIME

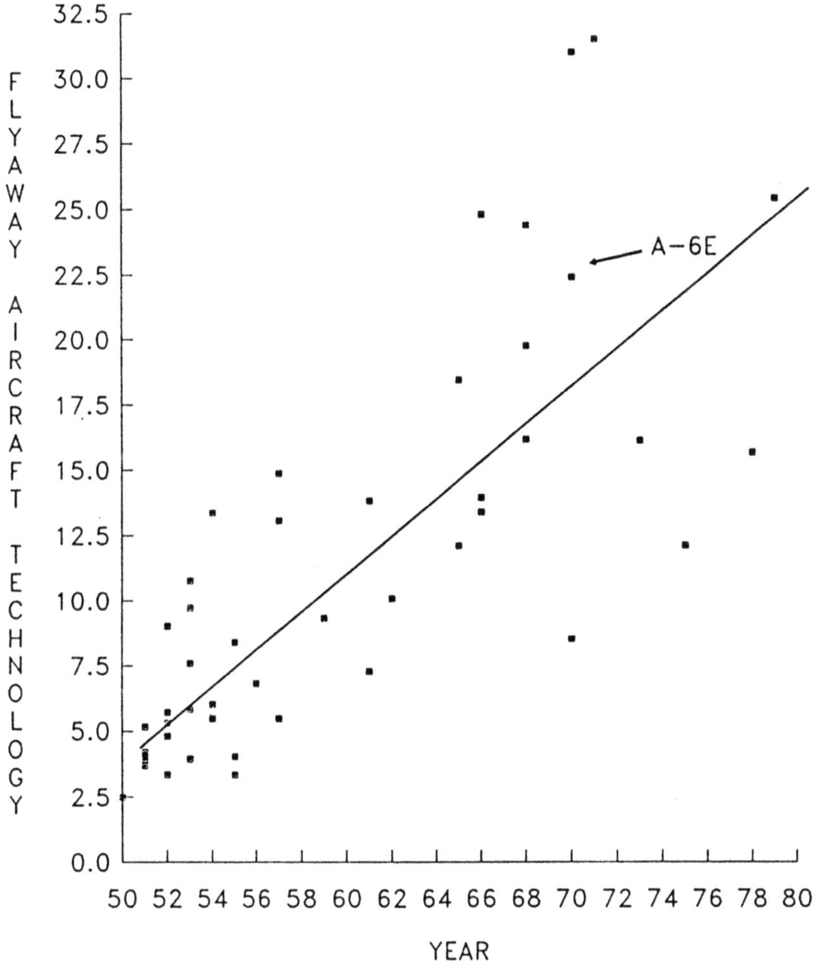

reflecting technological complexity or extension:

1. STAND: the average state-of-the-art of technology at the time of production of a system. (For any individual aircraft this is the predicted value from the trend line.)
2. ADVANCE: the extension in technology beyond the state-of-the-art. (For any individual aircraft, this is the residual from the regression model, or the deviation from the trend line.)
3. REACH: the total technology embodied in the system. (For any individual system this is simply STAND + ADVANCE.)[1]

Values for STAND, ADVANCE and REACH were determined for each of the three kinds of technology. Prefixes P, S. and F indicate PLATTECH, SYSTECH, and FLYTECH, respectively. (e.g., PADVANCE is the measure of advance for PLATTECH or platform technology.)

THE MEASUREMENT OF PRODUCTION COST

All cost data for the aircraft were taken from the US Military Aircraft Cost Handbook [3]. This section describes the steps taken to arrive at a production cost figure for each aircraft that could be considered comparable across the sample. Determination of comparable cost figures were hampered by three factors:

A. Costs were incurred at different points in time when the value of the dollar differed.
B. Aircraft were not purchased singly, but rather in "lots" of varying quantity.
C. Cost per unit tends to decline with additional units produced due to production "learning".

---

[1]Note, there are alternative ways of determining measures of STAND and ADVANCE. Rather than using a trend line to reflect the average state-of-the-art of technology, one could designate a specific individual system as a reference point. Candidates might be a) an immediate predecessor system or b) the predecessor system with the greatest REACH (maximum predecessor technology). The technology embodied in either reference system would constitute STAND, and ADVANCE would be measured as deviations from the specific reference system. (Of course the reference system would change as time progressed.) These alternatives were explored with no material enhancement of the analysis.

The raw data available consisted of costs and quantities per lot. The following procedures were employed to transform the available data into comparable cost figures.

1. All lot costs were converted to fiscal year 1981 dollars using Office of Assistant Secretary of Defense, Comptroller, composite price indices for major commodity procurement.
2. Cumulative quantities at the end of each lot were determined by summing the quantities in all preceding lots.
3. Cumulative average costs (FY81) at the end of each lot were determined by summing the costs of all preceding lots and dividing by the cumulative quantities.
4. Learning curves of the following form were fit to the quantity and cumulative average cost series:

$$C_Q = AQ^B$$

where

C_Q = Cumulative average cost for quantity Q
Q = Cumulative quantity
A = Cost of the first unit (estimated by the fitting procedure)
B = Constant (estimated by the fitting procedure)

5. The cumulative average cost of producing 100 units, CAC(100), was determined by setting Q at 100 and re-entering the learning curve to solve for C_Q.

This procedure is ad hoc but does provide a comparable average cost figure at a comparable quantity for all aircraft, taking into consideration the different learning rates experienced on different aircraft programs. The result is an average cost per unit of producing 100 aircraft.[2]

Cost data was available for three separate cost categories for each aircraft:

Airframe cost

---

[2]For discussions of learning curves and their relationship to production cost see Kaplan [14], Liao [18] and Womer [25]. For more detail on the specific procedures used to determine CAC see DePuy, et. al. [3]. The convention of determining CAC at 100 units has been adopted by other researchers. See, for example, Dodson [6].

Airframe plus engine cost
Total flyaway cost

The approach described above was applied to the three separate cost categories resulting in three variables to be used in the analysis:

FRAMCOST: CAC(100) for airframe cost.
PLATCOST: CAC(100) for aircraft platform (airframe & engine) cost.
FLYCOST: CAC(100) for flyaway aircraft cost.

Note that there is a direct correspondence between PLATCOST and the previously discussed PLATTECH measure, and between FLYCOST and FLYTECH. In these cases, the TECH variables measure technology and the COST variables measure cost for analogously defined components of the aircraft. FRAMCOST is a cost measure for airframes, but there are no corresponding TECH measures. (Without an engine the aircraft can't fly, so no separate measure of airframe performance or technology is possible.) Technology measures for platforms were used in attempting to explain airframe costs. Additionally, there are TECH measures for systems but no analogous cost measure. Technology measures for systems were used in some of the tests explaining FLYCOST, since the cost of avionics and weapons systems are included in the total flyaway cost.

Data to construct FRAMCOST and PLATCOST were unavailable for 11 aircraft. Data to construct FLYCOST were unavailable for nine. Those aircraft were deleted from further analysis. All cost variables are measured using millions of FY81 dollars.

PRODUCTION COST AND TECHNOLOGY: HYPOTHESES AND TESTS

The initial analysis concerns the association between production cost and the technology in the aircraft produced. Can technology measures, based on system characteristics, reliably predict production cost? The first hypothesis is that production cost increases with increases in the SOA of technology (the level of technological complexity). STAND reflects the average SOA of technology at the time of the commencement of production of an aircraft.

H_1: Production Cost = + f (STAND)

The second hypothesis is that production cost increases with the degree of technological extension of a program. ADVANCE captures this notion.[3]

H_2: Production Cost = + f (ADVANCE)

The third hypothesis follows from the mixed nature of the sample. The sample includes some aircraft which are the first series of a new design (eg. F-111A) and some which are follow-on series of an existing design (e.g. F-111B, F-111D, F-111F). It is reasonable to argue that sufficient production learning would occur during the first series of new design so that follow-on series would experience some reduction in cost. Hence

H_3: Production Cost = + f (first series of new design)

A dummy variable (SERIES) was created to capture this idea. SERIES was coded 1 for the first series of a new design and 0 for a follow-on series of an existing design.
Operationally, the hypotheses imply the following multiple regressions:

FRAMCOST = + f (PSTAND, PADVANCE, SERIES)
PLATCOST = + f (PSTAND, PADVANCE, SERIES)
FLYCOST = + f (FSTAND, FADVANCE, SERIES)

Following the recommendations of others (e.g. DePuy, et. al., [4]) regressions using both COST and ln(COST) measures as dependant variables were run. Using the natural log reduces the effect of extremes on the regression (particularly important when sample size is small). Additionally, regressions using COST as the dependant variable were found to be heteroscedatic (larger residuals at larger values of cost). This violates an assumption of

---

[3]REACH is a linear combination of STAND and ADVANCE and, hence, redundant for testing purposes; it contains no additional information.

regression that error variance is constant over all observations, resulting in residuals that are not of minimum variance. A common solution to this problem is to log the dependant variable (see Neter and Wasserman, [22]). Findings from using the two alternative measures were similar, but the use of ln(COST), produced higher R^2 values. Those results (models 1-3) are in Table 2.

TABLE 2
COST REGRESSIONS - ALL AIRCRAFT

Dependent Variable	Independent Variables	Coeff.	t	Prob.*	Model Statistics
Model 1:					
FRAMCOST	Intercept	-792	---	---	F = 21.30
	PSTAND	.206	6.17	.0001	Prob. = .0001
	PADVANCE	.212	5.11	.0001	R^2= .67
	SERIES	.363	2.07	.0233	Adj. R^2 = .64
Model 2:					
PLATCOST	Intercept	-.706	---	---	F = 27.71
	PSTAND	.219	7.36	.0001	Prob. = .0001
	PADVANCE	.198	5.35	.0001	R^2= .73
	SERIES	.388	2.48	.0094	Adj. R^2 = .70
Model 3:					
FLYCOST	Intercept	.321	---	---	F = 34.19
	FSTAND	.099	7.95	.0001	Prob. = .0001
	FADVANCE	.092	5.97	.0001	R^2= .75
	SERIES	.446	2.96	.0028	Adj. R^2 = .73
Model 4:					
FLYCOST	Intercept	.312	---	---	F = 39.94
	FSTAND	.104	9.84	.0001	Prob. = .0001
	PADVANCE	.189	6.48	.0001	R^2= .83
	SADVANCE	.589	2.36	.0122	Adj. R^2 = .81
	SERIES	.329	2.55	.0078	

* One tailed tests

All models in Table 2 are highly significant and explain a large proportion of the variance in production cost. All coefficients for the STAND, ADVANCE and SERIES predictors are also significant and positive, consistent with the hypotheses. The conclusion is that both the SOA of technology in general and the extension of technology in individual aircraft explain production cost. And the findings for the SERIES variable indicate an important "premium" in production cost for new designs.

Note that model 2 explains a greater proportion of PLATCOST than model 1 does for FRAMCOST. Since the two models contain the same predictor variables, this result is consistent with PSTAND and PADVANCE being surrogates for frame technology and measuring technology SOA and extension for airframes with "noise".

Model 4 in Table 2 is an alternative approach to explaining FLYCOST by using the separate ADVANCE measures for platform and systems, the two items making up the flyaway aircraft.[4] The basic conclusion to be drawn from model 4 is that additional explanatory ability is achieved by substituting PADVANCE and SADVANCE for FADVANCE.[5]

Tables 3 and 4 display analogous regressions for two subsamples: new design, first series and old design, follow-on series, respectively. In general, the findings are consistent with those from the full sample: R^2s are high and coefficients are positive and significant. Model and coefficient significance declines some from the full sample, which is to be expected given the smaller sample size in the subsamples.

There is one pattern of interest. For follow-on series (Table 4), coefficients for the ADVANCE predictors are larger than for the STAND predictors. (This is generally not the case in Table 3 for the new design aircraft.) The pattern becomes understandable by considering that new designs involve construction from the "ground up" of a new aircraft. Both achieving the current SOA (STAND) and extending it (ADVANCE) must be "paid for". A new series of an existing design, however, involves only "building from" an existing aircraft. Cost should then be more strongly driven by the extension to the existing aircraft that must be "paid for". In short, higher coefficients for ADVANCE for follow-on

---

[4]All STAND measures, being predicted values from a regression of TECH on time, are linear transformation of each other. Hence FSTAND is included in Model 4.

[5]Each model was also run using REACH in place of STAND and ADVANCE. R^2s decreased, but all regressions were highly significant, indicating that a measure reflecting total technology in systems does well as a substitute for the two separate measures reflecting technology trend plus extension.

TABLE 3
COST REGRESSIONS - NEW DESIGN / FIRST SERIES

Dependent Variable	Independent Variables	Coeff.	t	Prob.*	Model Statistics
Model 1:					
FRAMCOST	Intercept	-.647	---	---	F = 12.90
	PSTAND	.228	4.72	.0001	Prob. = .0004
	PADVANCE	.228	3.78	.0008	R^2= .60
					Adj. R^2 = .57
Model 2:					
PLATCOST	Intercept	-.509	---	---	F = 15.33
	PSTAND	.239	5.34	.0001	Prob. = .0002
	PADVANCE	.207	3.70	.0009	R^2= .64
					Adj. R^2 = .60
Model 3:					
FLYCOST	Intercept	.654	---	---	F = 21.87
	PSTAND	.108	6.17	.0001	Prob. = .0001
	FADVANCE	.101	4.31	.0002	R^2= .71
					Adj. R^2 = .68
Model 4:					
FLYCOST	Intercept	.426	---	---	F = 27.48
	PSTAND	.121	8.45	.0001	Prob. = .0001
	PADVANCE	.198	5.08	.0001	R^2= .83
	SADVANCE	.874	3.02	.0039	Adj. R^2 = .80

* One tailed tests

series is a plausible result, and suggests that the ADVANCE and STAND measures do meaningfully capture elements of importance in explaining the production costs.

PRODUCTION COST VARIANCES

Predictions for production cost, given the technology embodied in the aircraft, can be created by taking the predicted values from the Table 2 regressions (models 1, 2, and 4) and converting (un-logging) to arrive at estimated production cost. Actual costs of course differ from the estimated costs. Variances were constructed by subtracting (actual - estimated), which can be interpreted as cost over(under)runs, given the technology produced. These cost variance measures are, of course, not measures of cost overruns or underruns in the most traditional sense of being measured relative to a budget. Traditional

TABLE 4
COST REGRESSIONS - OLD DESIGN / NEW SERIES

Dependent Variable	Independent Variables	Coeff.	t	Prob.*	Model Statistics
Model 1:					
FRAMCOST	Intercept	-.499	---	---	F = 12.64
	PSTAND	.174	3.40	.0027	Prob. = .0011
	PADVANCE	.209	3.31	.0031	R^2= .68
					Adj. R^2 = .62
Model 2:					
PLATCOST	Intercept	-.398	---	---	F = 19.93
	PSTAND	.186	4.44	.0004	Prob. = .0002
	PADVANCE	.206	3.98	.0009	R^2= .77
					Adj. R^2 = .73
Model 3:					
FLYCOST	Intercept	.466	---	---	F = 23.14
	FSTAND	.087	4.20	.0005	Prob. = .0001
	FADVANCE	.091	3.93	.0008	R^2= .77
					Adj. R^2 = .73
Model 4:					
FLYCOST	Intercept	.495	---	---	F = 24.86
	PSTAND	.089	4.83	.0002	Prob. = .0001
	PADVANCE	.222	4.82	.0002	R^2= .85
	SADVANCE	.021	.04	.4849	Adj. R^2 = .82

* One tailed tests

variance measures most frequently compare resource inputs (costs) relative to budgeted inputs. The variance measures here compare actual costs with expected costs based on output, where output is measured by the technological performance of the aircraft. The term "variance" is used here in the accounting sense (deviation from a target cost), not in the statistical sence; they are not measures of cost dispersion.

Plots of the cost variances, arranged by REACH of the flyaway aircraft are in Figures 4 through 6. Two aircraft are highlighted in the figures, the F/A-18A and the F-14A. Figure 6 shows a large positive variance (cost overrun) for flyaway aircraft cost was incurred on the F/A-18A. Figures 4 and 5 show that the F/A-18A also experienced the largest positive variances on airframe and platform costs, suggesting that these two cost elements contributed greatly to the expensive flyaway cost. Figure 6 shows the largest negative variance was experienced on the F-

FIGURE 4

PLOT OF FRAME COST VARIANCE VERSUS FLYAWAY REACH

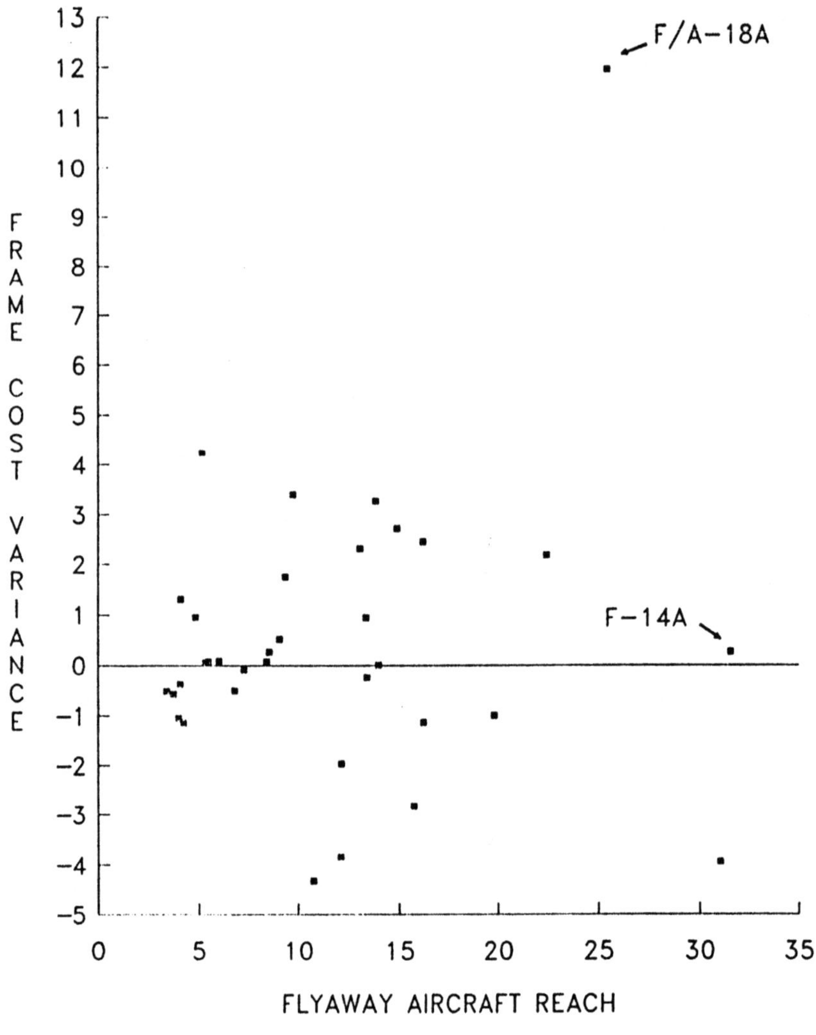

FIGURE 5
PLOT OF PLATFORM COST VARIANCE VERSUS FLYAWAY REACH

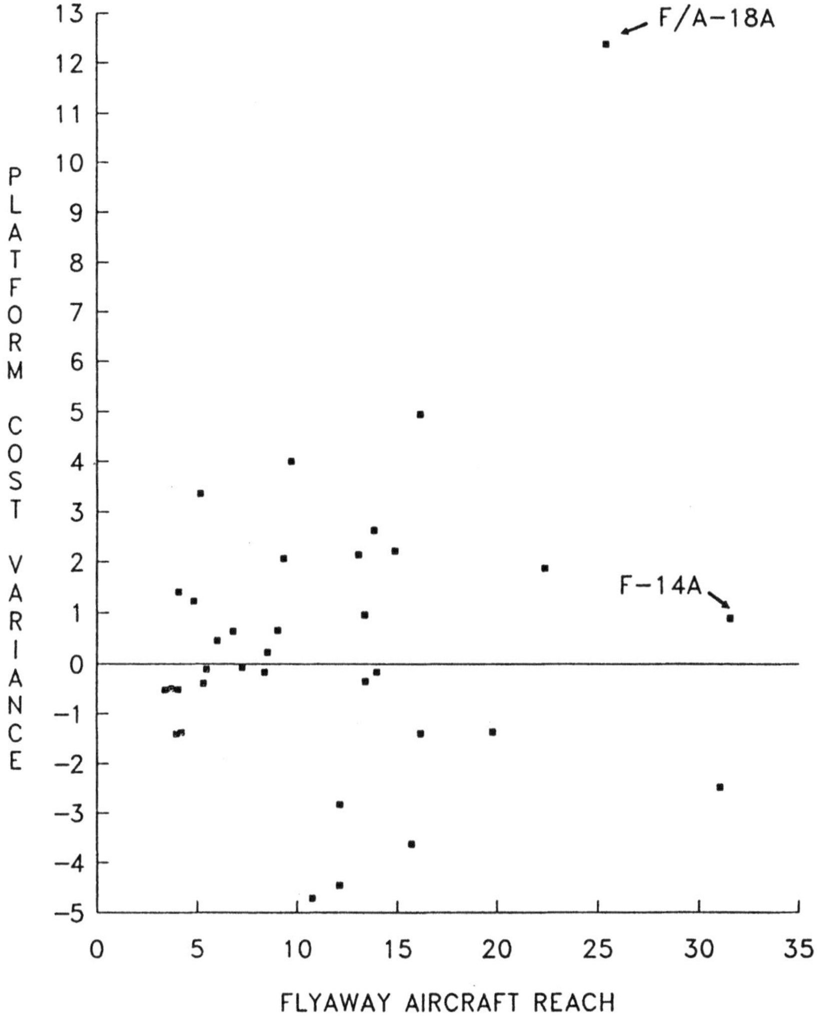

FIGURE 6

PLOT OF FLYAWAY COST VARIANCE VERSUS FLYAWAY REACH

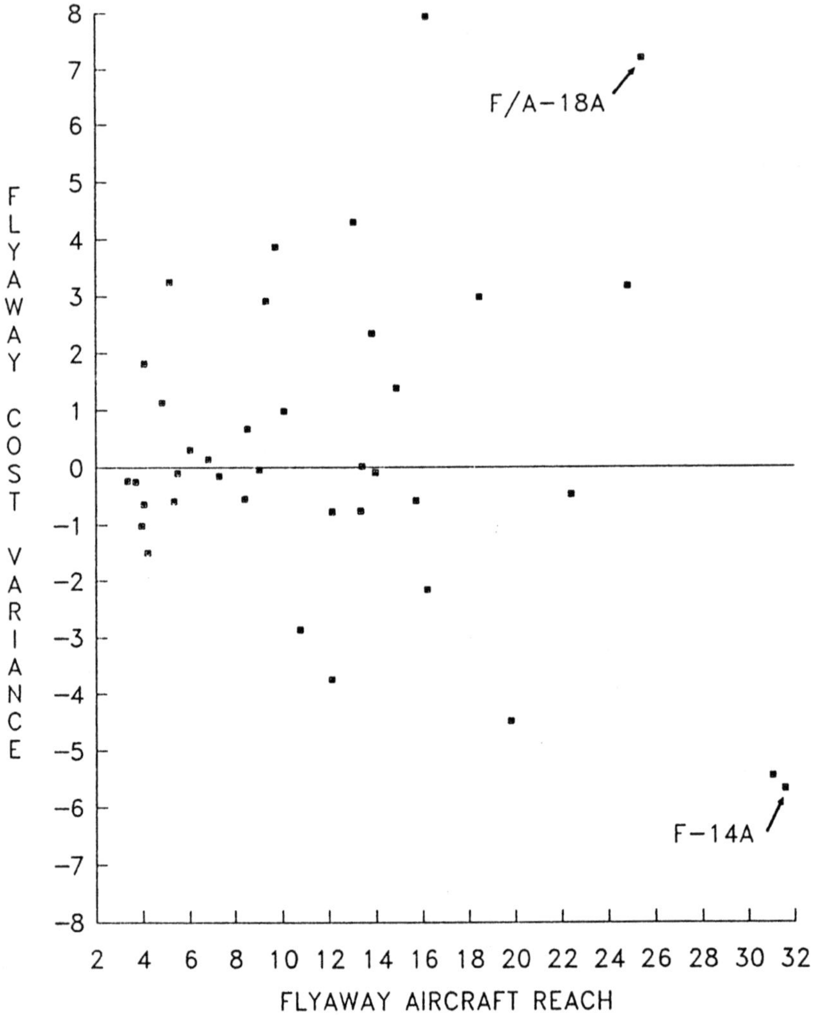

14A; it was inexpensive relative to the technology embodied in it. The large "savings" was apparently not due to an inexpensive airframe or platform; variances in Figures 4 and 5 are close to zero. This suggests that the avionics and weapons systems added to the platform were cost effective. They enhanced the flyaway performance of the aircraft substantially relative to their additional cost.

The variances may be interpreted as measures of cost overruns or cost savings, relative to the technology embodied in the systems. The following sections address the question of possible causes of the variances.

EXPLAINING COST VARIANCES - ENVIRONMENT FACTORS

Clearly defense procurement, particularly for major weapon systems, is specialized in nature. Both the product and market are not typical of products and markets in general. Major weapon systems are large dollar items which may represent a substantial segment of a manufacturer's business. Major weapons systems incorporate significant innovation with state-of-the-art hardware and substantial uncertainty in development. The market for defense systems is unusual, with a single (monopsonistic) buyer and usually only a few (oligopolistic) sellers. Pricing strategy for such items is likely to be an important strategic decision.

Prices are determined primarily through a bid and negotiation process. A bid is accepted and a contract for a specified number of units is negotiated prior to production. Prices (costs to the government) are specified in the contract and are based on costs incurred ("cost plus") using some agreed upon formula. Cost estimates and their source are disclosed at the time of contract negotiation, so some agreement on the validity of cost estimates is established up front.

When does the government pay "too much" or "too little" for high technology systems? Or phrased alternatively, what conditions are associated with positive or negative cost variances, given the technology embodied in the system? Several factors might influence the price that would be offered by the contractor, and accepted by the government, and consequently have some impact on program costs. The factors fall into three broad areas:

1) program characteristics, 2) the political environment, and 3) the economic environment.

Several variables are discussed below. Each is an attempt to reflect some feature of a program or the procurement environment existing at the time of program initiation. For each factor, how that factor might influence the prices that are offered by contractors and accepted by DoD are discussed. To the extent to which these factors influence prices paid, they provide possible explanations for cost variances experienced.

Program Value. Larger programs may be associated with greater risk to a contractor. If a program is "small", experiencing unexpected costs or losses on the program, while damaging to a firm, would likely not be critical. In contrast, unfavorable performance on a "large" program could have significant implications for the performance of the firm as a whole. Greater down-side financial risk exists. Additionally, larger programs may, because of their size and complexity, be more difficult to manage and control. Greater managerial risk exists. Consequently, it was expected that, as the size of a program (as measured by the dollar value of the program to the contractor) increased, contractors would seek, and be allowed, a "premium" to compensate for additional risk.[6] Hence, program value is hypothesized to be positively associated with cost variance.

H_4: Cost Variance = +f (Program Value)

Number of Lots. Features of the acquisitions environment preclude the use of a single, unchangeable contract covering all units to be manufactured during a weapon system acquisition program. Due to the complex nature and state-of-the-art technology involved in major weapon systems, contracts may be revised to accommodate design and production changes. Additionally, because of the nature of the federal budget process, funding for units procured under a weapon system program is

---

[6]Regulations governing U.S. DoD procurement under cost plus type contracts specifically authorize increased profit to the contractor (resulting in higher cost to DoD) to compensate both for higher contractor risk and greater utilization of contractor facilities. (See U.S. DoD Federal Acquisitions Regulations Supplement, Part 215.)

reviewed and approved on an annual basis. The result is that system procurement typically occurs in stages under different contracts, each covering the acquisition of a distinct "lot", consisting of a subset of the total number of units produced. Contractors frequently "buy-in" to a program with a low bid for the initial lot contract, and attempt to generate a satisfactory return by negotiating more favorable prices on subsequent lots once their position as the manufacturer has been established.[7] It was expected that a contractor's ability to increase price would be associated with the number of opportunities for negotiating additional contracts. Consequently, a positive relationship between the number of lots in a procurement program and cost variance was hypothesized.

H_5: Cost Variance = +f (Number of Lots)

<u>Defense Spending</u>. What was the political and budgetary environment like at the time a program was initiated? Were constraints being imposed on defense spending? Were defense or non-defense programs favored? It was felt that contractors would have less incentive to offer a low price (and perhaps government negotiators would have less pressure on them to demand a low price) if the political environment appeared favorable to defense spending.[8] A positive relationship between cost variance and the

---

[7]Buying-in with an initial low bid is cited by researchers as a common cause of cost growth on government contracts. The ability of a contractor to increase price after its position as the manufacturer has been established is reduced if a second source manufacturer can be set up. But problems related to technology transfer, the existence of proprietary information and the high cost of second source qualification and start up prohibit second sourcing for the types of systems studied here. The impracticality of second sourcing places the sole source manufacturer in a strong negotiating position on subsequent production lots. (See White and Hendrix, [24], p. 63 and p. 93.)

[8]Evidence from research on pricing strategy in the aerospace industry [21] supports the conclusion that, as defense spending increases, contractors adopt strategies that tend toward higher initial prices for aircraft system.

degree of defense spending (as a proportion of federal spending) at the time of program initiation was hypothesized.

H_6: Cost Variance = +f (Defense Spending)

Program Funding. There is inevitably some uncertainty concerning the long run commitment of the government to individual weapon systems. Long run plans may be made, but the federal budget is discussed and revised annually. Programs that are supported one year by an administration or congress may be cut in subsequent years as the administration, congress or political conditions change. To the extent that long run commitment to a particular weapon system is doubtful, contractors may perceive greater risk and demand a higher price. If commitment to a program is not in doubt, contractors may have greater confidence that program curtailment will not threaten returns and, consequently, offer a lower price, consistent with the lower risk. Commitment to a program is not readily measured, but funds allocated to a program, as reflected in annual obligational authority, may provide an indication of the government's willingness to commit to a program. "Early" allocation of funds may reflect a strong initial commitment. The initial year obligational authority for a program was divided by the total obligational authority over the life of a program to create a measure reflecting the proportion of the project that was funded "up front".[9] This measure of early funding was expected to be negatively associated with cost variance.

H_7: Cost Variance = -f (Program Funding)

Presidential Party. The conventional wisdom concerning the views of the two major U.S. political parties toward defense spending considers Republicans (Democrats) to be biased toward devoting resources to defense (social) programs. A more favorable climate for defense spending may encourage contractors to seek, and the government to accept, higher prices for defense

---

[9]Research by Moses [21] demonstrates a significant relationship between the degree of initial year funding for weapon systems programs and the adoption of low initial price pricing strategies by contractors.

programs. While congress has been generally controlled by Democrats during the period of this study, the Presidency has changed hands several times. A positive association between a Republican presidency at the start of a program and cost variance was expected.

H_8: Cost Variance = +f (Republican Party)

Capacity Utilization. High capacity utilization,[10] ceteris paribus, should typically be associated with a greater number of active projects for a firm and a greater volume of activity. Because of this, two effects may occur. First, fixed capacity and corporate overhead costs may be spread over the larger number of projects, resulting in a relatively lower cost per project.[11] Second, risks associated with a single project may be offset by risks on other projects. This benefit from the offsetting of risks (i.e., a portfolio effect) may permit a contractor to accept a relatively lower price on a specific project. Given one or both of these effects, the degree of capacity utilization was hypothesized to be negatively associated with cost variance.

H_9: Cost Variance = -f (Capacity Utilization)

Inflation. Inflation makes future dollars worth less than current dollars. When the inflation rate is high contractors may compensate for its effect by building a cushion into the price they offer in order to cover expected higher costs. If this effect takes

---

[10]Capacity utilization was measured for the aerospace industry as whole, rather than for individual firms. Work by Greer and Liao [11] shows that industry capacity utilization is a better predictor of firm pricing and bid behavior than is firm specific capacity utilization. This result holds because, in a competitive industry, individual firm actions are influenced by the actions of competitors such that the "average" capacity utilization of the industry appears to drive behavior.

[11]Consistent with this hypothesis, Greer and Liao [12] demonstrate that unit costs are inversely associated with industry capacity utilization in the aerospace industry, when contracts are sole source, cost plus type.

place, high prices may occur when inflation rates are high. A positive association between the rate of inflation at program start and cost variance was hypothesized.[12]

H_{10}: Cost Variance = +f (Inflation Rate)

General Economic Conditions: Economic conditions - growth or contraction - may influence program cost. If the economy is robust, demand for products should be relatively greater, opportunities for alternative commercial projects supplied by contractors may be more plentiful, and incentives to compete on price for a particular defense contract may be reduced. When economic contraction occurs, new defense programs may appear more appealing and the increased incentives to compete for such contracts may result in lower prices. A positive relationship between the rate of GNP growth at the time of program start and cost variance experienced on a program was hypothesized.

H_{11}: Cost Variance = +f (GNP Growth)

Table 5 summarizes the explanatory variables and their measurement.

ANALYSIS OF EXPLANATORY VARIABLES

To test the hypotheses that program and environment factors affect the costs incurred to acquire high-technology weapon systems, three multiple regression models were constructed. The cost variances (FRAMVAR, PLATVAR, FLYVAR), representing the portion of cost that could not be explained by the technology in the aircraft, were regressed on the set of explanatory variables. Results are contained in Table 6.

Several findings are evident from the regressions. First, all three

---

[12] Lehman [17] argues that the Program, Planning and Budget System builds past inflation into future price estimates; that contractors, aware of the upward bias caused by the built-in inflation factors, automatically raise prices to the level they know is permitted by the inflation factor; and that this process guarantees price escalation. His discussion centers on the acquisition of the F/A-18A.

TABLE 5
PROGRAM AND ENVIRONMENT EXPLANATORY VARIABLES

Program Variables	
1. Program Value:	Average annual dollar value of a program over the program's life. Measured in 1981 dollars (millions).
2. Number of lots:	Total number of individual lots contracted for over a program's life.
Political Variables	
3. Defense Spending:	Defense spending as a percent of total federal spending. Measured at time of program start.
4. Program Funding:	Initial year obligational authority divided by total obligational authority over the life of a program.
5. Political Party:	Presidential party in power at time of program start. (Republican = 1, Democrat = 0.)
Economic Variables	
6. Capacity Util.:	Percentage capacity utilization of aerospace industry at year of program start.
7. Inflation:	Percentage change in Producer Price Index-Industrial at year of program start.
8. Economic Growth:	Percentage change in price-adjusted GNP at year of program start.

models are significant and explain a fair proportion of the variance in the dependent variable. Second, all of the eight predictors (except Presidential Party) have significant coefficients in one or more of the models, and, when significant, the coefficient signs are as hypothesized.

The strongest results, in terms of level of significance and consistency across the three models, are for Program Value, Defense Spending and Inflation; these factors are associated with all three cost variance measures.[13] The conclusions follow from

---

[13]Of course the tests are not independent. Since flyaway aircraft cost includes platform cost, and platform cost includes airframe cost, the cost measures (and cost variance measures) are interrelated.

TABLE 6
COST VARIANCE REGRESSIONS - ENVIRONMENT FACTORS

Dependent Variable:	FRAMVAR		PLATVAR		FLYVAR	
Explanatory Variables	Coeff.	t	Coeff.	t	Coeff.	t
Intercept	-15.0	----	-17.4	---	-27.6	----
Program Value	.0080	4.05***	.0082	3.90***	.0051	3.02***
Number of lots	.3274	1.81**	.3212	1.66*	-.134	-.65
Defense Spending	17.23	3.60***	15.56	3.42***	9.05	1.66*
Program Funding	-2.50	-.88	-2.48	-.81	-4.43	-1.34*
Pres. Party	-1.02	-1.17	-.49	-.52	.77	.74
Capacity Util.	-.072	-1.57*	-.076	-1.53*	.0038	.07
Inflation	63.99	4.12***	67.19	4.04***	35.56	2.04**
Economic Growth	6.12	.49	8.02	.60	19.67	1.34*
Model Statistics						
F =	5.02		4.80		2.86	
Prob. =	.0008		.0010		.0179	
R^2 =	.61		.60		.44	
Adj R_2 =	.49		.47		.29	
n =	35		35		38	

* Significant at probability < .10, one tailed tests
** Significant at probability < .05, one tailed tests
*** Significant at probability < .01, one tailed tests

the hypotheses: Larger programs, perhaps because they are more risky or more difficult to manage, tend to be associated with cost overruns. When defense spending is high, cost overruns tend to result. This is consistent with an environment favorable to defense spending leading to acceptance of a higher price by DoD. Cost overruns also tend to follow periods of rapid inflation. As suggested by Lehman [17], this may be due to an institutionalized planning and pricing system that builds past inflation rates into future cost estimates.

The Capacity Utilization variable is most significant for FRAMVAR and least significant (not significant) for FLYVAR. This pattern is perhaps understandable. Recall that the three costs of interest are progressively more comprehensive measures:

Airframe Cost

+ Engine Cost
= Platform Cost
+ Avionics and Weapons Systems Cost
+ Miscellaneous Cost
= Flyaway Cost

The prime contractor for an aircraft program will be in the aerospace industry. So the aerospace industry capacity utilization measure may be most directly related to the actions of the prime contractor. The prime contractor typically constructs the airframe, but subcontracts out engine and systems manufacture. A high association between aerospace capacity utilization and the costs directly related to the prime contractor's manufacturing efforts (i.e., airframe costs) may not be surprising. On the other extreme, total flyaway cost includes systems, which are typically subcontracted to firms in the electronics industry. Consequently, flyaway aircraft cost should be (relatively) less affected by conditions in the aerospace industry. Thus, the lower association of FLYVAR with aerospace industry capacity utilization may be understandable.

The number of lots variable is also most significant in the FRAMVAR regression and least (not) in the FLYVAR regression. A somewhat analogous explanation may apply. The government contracts with the prime contractor for specific lots. Hence, number of lots is found to be associated with what the prime contractor manufactures (airframes). Arrangements between the prime contractor and subcontractors to acquire electronic systems may be only indirectly influenced by the number of lots. Hence, flyaway cost, which includes the electronic systems cost, is found to be unrelated to the number of lots. In short, capacity utilization and number of lots may be expected to more strongly influence prime contractor actions and the strongest (weakest) results are found when explaining costs most (least) directly under prime contractor control.

Broadly, the overall findings are consistent with the identified factors (except Presidential Party) influencing costs as expected. Aspects of the program, political environment and economic environment do reflect conditions indicating when cost overruns or savings may be expected.

CONCLUSIONS

This study demonstrated the use of technology measurement procedures in transforming descriptions of multiple properties of systems into two technology measures: STAND, representing the average state-of-the-art of technology at the time of production of a system; and ADVANCE, representing the extension in technology for a system beyond the state of the art. For three categories of cost, these two technology measures were able to explain from 67% to 83% of the variance in cost across the systems studied. Hypotheses concerning the relationship between these measures and cost were confirmed. Additionally, the strength of the relationship between these two measures and cost differed between two subsamples in a manner consistent with the intuitive meaning of the measures.

Recent unpublished research [10, 13] used alternative technology measurement procedures from those used in this study to create conceptually analogous measures of STAND and ADVANCE, and also found those measures useful in predicting cost in different samples of high-technology systems.[14] A principle conclusion is that filtering multiple properties of systems through a technology measurement process to create simple, intuitively understandable, summary measures of technology state-of-the-art and technology extension results in measures that contain considerable information useful in predicting cost. This suggests that technology measurement procedures existing in the literature [1,2,5,7,9,10,20] provide a practical basis for creating cost-estimating relationships for high-technology systems.

This study also documented that actual cost of systems are not solely predictable with reference to the technology contained in the systems. Environmental factors play a predictable role. Hypotheses concerning the association between environmental variables and cost were confirmed (in at least one test) for seven of eight factors identified in the study. And models containing these factors were able to explain 44% to 61% of cost variance. Measurement of the environmental variables was somewhat crude.

---

[14]Greer [10] used a ellipsoid approach developed by Dodson and Graver [5], while Greer and Moses [13] used a linear time regression. Both studies investigated the links between technology measures and the development cost of satellite systems.

And many of the factors addressed in this study are unlikely to be generalizable to situations different from the defense contracting arena from which the sample of high-technology systems was taken. However, establishing that environmental factors do have a predictable effect on cost provides a first step toward future research that may consider incorporation of environmental variables into investigations of the cost of extending technology.

REFERENCES

1. Alexander, A. J., and B. Mitchell, "Measuring Technological Change of Heterogeneous Products," Technological Forecasting and Social Change, 27 (1985), pp. 161-195.
2. Alexander, A. J., and J. R. Nelson, "Measuring Technological Change: Aircraft Turbine Engines, "Technological Forecasting and Social Change, 5 (1973), pp. 189-203.
3. DePuy, W., R. Moyer, P. Palmer, B. McKinney, G. Kreisel, S. Balut and G. McNichols, U.S. Military Aircraft Cost Handbook, # TR-8203-1, (Management Consulting and Research, Inc., Falls Church, Virginia, 1 March 1983).
4. DePuy, W., R. Moyer, J. Ford and G. McNichols, Task IA Report: Feasibility Study for Integrating a Cost Methodology into the TASCFORM - Air Model, # TR-8007-1R, (Management Consulting and Research, Inc., Falls Church, Virginia, 18 November 1980).
5. Dodson, E. N., and C. A. Graver, "An Approach to Quantitative Measurement of Advances in State-of-the-Art", Internal Memorandum (Releasable) IMR-997, General Research Corporation (Santa Barbara, 1969).
6. Dodson, E. N., "Technological Change and Cost Analysis of High-Technology Systems", IEEE Transactions on Engineering Management (May 1977), pp. 38-45.
7. ________, "Measurement of State-of-the-Art and Technological Advance", Technological Forecasting and Social Change, 27 (1985), pp. 129-146.
8. Franklin, D., "Determination of Substitutes for Capacity Utilization Rates for Individual Companies in the Aerospace Industry", Unpublished Masters Thesis, Naval Postgraduate School, Monterey, CA, June 1984.
9. Gordon, T. J., and T. R. Munson, "A Proposed Convention

for Measuring the State-of-the-Art of Products or Processes", Technological Forecasting and Social Change, 20 (1981), pp. 1-26.

10. Greer, W., A Method for Estimating and Controlling the Cost of Extending Technology, Naval Postgraduate School Technical Report, # NPS-54-88-002, (Monterey, CA, 1988).
11. Greer, W. and S. Liao, Cost Analysis for Competitive Major Weapon Systems Procurement: Further Refinement and Extension, Naval Postgraduate School Technical Report, # NPS-54-84-023, (Monterey, CA, 1984).
12. Greer, W. and S. Liao, "Weapon Pricing Models for Defense Acquisition Policy", Journal of Accounting and Public Policy, 6, 1987, pp. 271-284.
13. Greer, W. and O. Moses, "Estimating and Controlling the Cost of Extending Technology", Unpublished Manuscript, University of Iowa, April 1989.
14. Kaplan, R., Advanced Management Accounting, (Prentice Hall, 1982), pp. 97-105.
15. Keeney, R. and H. Raiffa, Decisions with Multiple Objectives: Preferences and Value Tradeoffs, (John Wiley & Sons, 1976).
16. Knight, K. E., "A Functional and Structural Measurement of Technology",Technological Forecasting and Social Change, 27 (1985), pp. 107-127.
17. Lehman, J., Command of the Seas, (Charles Scribner & Sons, New York, 1988), Chapter 7.
18. Liao, S., "The Learning Curve: Wright's Model vs. Crawford's Model". Issues in Accounting Education, Vol 3, No. 2, Fall 1988, pp. 302-315.
19. Lienhard, J. H., "The Rate of Technological Improvement Before and After the 1830's", Technology and Culture (July 1979), pp. 515-530.
20. Martino, J. P., "Measurement of Technology Using Tradeoff Surfaces". Technological Forecasting and Social Change, 27 (1985), pp. 147-160.
21. Moses, O., "Determinants of Contractor Pricing Strategy", Program Manager, Forthcoming.
22. Neter, J. and W. Wasserman, Applied Linear Statistical Models, (Irwin, 1974).
23. Timperlake, E., W. Vogt, and L. Sullivan, The Tascform -

Air Model, Analytic Sciences Corporation Research Report, # TR-1334-3 (Arlington, VA, February 1980).

24. White, E. and V. Hendrix, Defense Acquisition and Logistics Management, (National Defense University, Washington D.C., 1984).

25. Womer, N. K., "Learning Curves, Production Rate and Program Cost", Management Science, April 25, 1979, pp. 312-319.

II. Lot Sizing and Cost Control

Allocated Cost Structures to Control Equipment Usage "Bottlenecks"

J. Christopher Westland
University of Southern California, Los Angeles, CA

ABSTRACT

Equipment resource "bottlenecks" arise where demand exceeds a resource's ability to provide service. Cost allocation systems which control bottlenecks must reconciles three, often incompatible, objectives – Pareto efficiency, incentive compatibility and information efficiency. This research developes such a cost allocation system where there exists a positive cost of waiting for a resource such as a critical piece of equipment, and a positive benefit to using that resource. This costing approach should be useful in determining internal transfer prices for factory machines or departments whose services must be shared by different operations, and which in the absence of discouragement from a significantly high allocated cost, would tend to be overloaded. This research shows that commonly used cost allocation mechanisms cause resource consumers to underconsume resources, and thus lower efficiency. These problems can compound themselves by misleading management into suboptimal use of equipment in the short run, as well as incorrect acquisition of equipment in the long run.

1: Motivation and Objectives for this Research

Potential for economies of scale in manufacturing can best be realized where expensive equipment is shared to the maximum possible extent. In capital intensive industries, competitive advantages accrue to production facilities with large capital investments, which can achieve an optimal mix of production with long runs, short setup times, level usage and few "bottlenecks" in usage in costly equipment. The efficient and effective sharing of equipment is often perceived as a scheduling problem. Risk averse management sets as the primary objective in scheduling the avoidance of usage bottlenecks. This research provides an alternative to scheduling algorithms which is informationally efficient, and attains constant usage for expensive, shared equipment.

Equipment use scheduling for factory production involves a considerable amount of uncertainty and expediting, much of this arising from uncertainty about timing and quantity of needs of users of the equipment. Often the usage schedule of an expensive, and heavily used piece of equipment cannot be effectively preordained. Individual users of the equipment will find the resource less desirable when bottlenecks are common, because waiting, breakdowns and so forth may become highly dysfunctional.

This research explores cost allocation schemes which effectively reduce usage uncertainty, and allow equipment usage to be smoothed and leveled in the face of uncertainties of user demand. The research evolves by exploring the definition of usage levels for each user, in a way that will be shown below to be optimal; then a cost allocation structure is chosen that will allow equipment users to choose the optimal usage level, in a simplified setting involving one type of raw material, and one type of processed output; this setting can be generalized to include multiple factors of production. The research develops a cost allocation scheme which meets three desirable objectives.

(Objective 1) The cost allocation scheme insures that no one user can be made better off without making another user worse off based on usage and price for usage – i.e. the scheme motivates *Pareto optimal* usage. Where a large number of users are concerned, and/or where lines of authority are equivocal, and/or where conflict between users already exists, Pareto optimal usage levels will minimize and ameliorate problems arising from conflicting claims on scarce equipment resources.

(Objective 2) The cost allocation scheme provides an approach to user acquisition of equipment time that is *incentive compatible* – i.e. it is to each individual user's benefit to use the equipment to the extent defined prior to calculating costs, and thus to the extent most beneficial to the organization. Note that a cost allocation defined in this manner allows management to impose its own objectives on machine users. Additionally, maintenance schedules, technical upgrades, research and development of processes and other non-production uses of the machine may be included in the machine usage schedule in an incentive compatible manner. Current cost allocation and transfer pricing mechanisms completely ignore non-production usage of the machine (except peripherally and after the fact in "idle" time calculations).

(Objective 3) The cost allocation scheme assumes that different users are generally unaware of each other's schedules for machine usage, and select usage levels without reacting to the effect of their usage on the usage of others. Although some exchange of information may actually accrue, it is unrealistic to assume that every user has the time or resources to be aware of every other user's schedule. In practice, the exchange of such information is likely to be minimal.

When goods are shared by several people, these three objectives are generally incompatible. Objective (3), combined with the assumption that usage is motivated by a cost allocation scheme, provides a type of Walrasian mechanism. Hurwicz [1977], Walker [1977] and Osana [1978] have shown Walrasian equilibria to be informationally efficient – i.e. they minimize the amount of information which must be exchanged between equipment users to attain objectives (1) and (2). But, when the good in question is shared between many users, such as expensive production machinery, objective (3) leads to activities, in equilibrium, which are generally inconsistent with objectives (1) and (2). The cost allocation scheme developed below reconciles objective (3) with objectives (1) and (2) for shared equipment.

Objective (2) is generally inconsistent with (1) – i.e. without outside incentives, the equipment user's actions will not be the actions which are best for the organization. Equipment users will tend to *suboptimize*, putting their own needs prior to that of the company as a whole. Within the scope of organizational coordination structures, prior research, e.g. Jensen and Meckling [1976], suggests that individual end-users should align their personal goals with those of the organization, and organizational coordinating structures should support this goal congruence. Conventional wisdom supports a similar concept, the "yes-man" concept, which asserts that the personal goals of effective employees should concur with those of the organization a high percentage of the time. Congruence of personal goals with corporate goals, e.g. profitability, growth, and survival, has been promoted as desirable attributes of cost allocation and performance evaluation systems by Vancil [1978] and others.

Pareto optimality reflects the "yes-man" concept for a particular class of group utilities – the *Samuelson-Bergson* group utility functions. Samuelson (1954) defined this group utility function $\mathfrak{F}=\mathfrak{F}(\Pi_1(\cdot),...,\Pi_m(\cdot))$ which, when $\partial\mathfrak{F}/\partial\Pi_i(\cdot)>0$ $\forall$i, satisfies the Pareto criterion, where $\mathfrak{F}$ reflects a corporate goal, and Π_i is a user goal.

In this research, Π_i is defined to be the revenue generating ability of user i from selling the finished good in some market, or using it as a subassembly in another good, although the analysis allows a more general set of measurements or utilities to be used. Π_i measures the value of the finished good in a particular market, and is approximated by accounting revenue. $\mathfrak{F}=\mathfrak{F}(\Pi_1(\cdot),...,\Pi_m(\cdot))$ is consistently defined to be the revenue generating capability of the organization. Suboptimization and internal transfers of goods imply that $\mathfrak{F}(\Pi_1(\cdot),...,\Pi_m(\cdot))\leq\sum_{i=1}^{m}\Pi_i$ (e.g. see Hage, Aiken and Marrett [1971]; Kochen and Deutsch [1980]; Malone and Smith [1988]).

Where the organization has reward mechanisms for promoting goal congruence, a Samuelson-Bergson group utility functions $\mathfrak{F}(\Pi_1(\cdot),...,U_m(\cdot))$ will reflect the organizational *reward/penalty structure* set in place by management to enforce goal congruence. Samuelson-Bergson group utility functions reconcile objective (1) with objective (2).

2: Optimal Equipment Usage

In the short run, the total amount of equipment is invariant, and this leads to capacity constraints. The service provided by the equipment hardware gradually degrades as more users are serviced, until no new users may be added and capacity is reached. Additionally, non-productive uses of equipment involved in maintenance and so forth, are required in increasing amounts as machinery becomes more complex and costly.

Use of the equipment resource for processing increases the utility of the raw material input by the users. The simplified analysis considered here assumes two types of material – raw material, and processed material. The utility of processed material is that of raw material, *plus* an increment representing the cost of the equipment usage required to transform the raw material to a finished good. The user of equipment has the alternatives of either not processing raw material, or of "outsourcing" raw material if the cost for using this equipment is too high.

Assume that the organization enforces chosen usage levels for the m users through reward/penalty mechanisms for promoting goal congruence, reflected in this model via a Samuelson-Bergson group utility function $\mathcal{F}(\Pi_1(\cdot),\ldots,\Pi_m(\cdot))$. Let the raw materials input to the equipment be designated u_1. Let the processing which produces the transformed output be designated u_2; u_2 may interchangeably be perceived as the product of this process or the processing that produces the product. Users have positive increasing willingness to pay for both raw material and material processed by the machine. This implies that more of either resource will improve the user's revenue generating capability $\partial\Pi_i/\partial u_{i2}>0$ and $\partial\Pi_i/\partial u_{i1}>0$, where u_{ij} is user i's consumption of resource j where resource 1 is raw input, and resource 2 is processed output.

Assume that input quantity u_1, output quantity u_2 and capacity ω are measured in arbitrary units. For a machine with m users, let the ith user's revenue function be $\Pi_i(u_{i1}, u_{i2}, u_2)$ where $u_2=\sum_{i=1}^{m}u_{i2}$ is the total usage of equipment processing. In this case user i's preferences for processing depends upon j's preferences for processing and total corporate revenue decreases with higher total equipment usage; $\partial\Pi_i\ /\partial u_2 < 0$.

Let the equipment performance allow transformation of the total amount of raw material u_1 into the total amount of processed material u_2, with out-of-pocket costs summarized in an implicit cost function $C(u_1, u_2)=0$. Assume that there exists some process for feeding back to the equipment users the production levels required to fulfil their processing needs. Assume no wastage; this constraint can be relaxed without problems, but makes the calculation more complex. Then each equipment user's revenue maximization problem with respect to scheduling usage of the equipment is

$$\max_{u_{i1},\, u_{i2}} \quad \Pi_i(u_{i1}, u_{i2}, u_2) \qquad (1)$$

subject to $C(u_1, u_2)=0$ (2)

The Lagrangian is

$$\Pi_i(u_{i1}, u_{i2}, u_2) - \lambda_1 C(u_1, u_2) \tag{3}$$

with first order conditions

$$\frac{\partial \Pi_j}{\partial u_{i1}} - \lambda_1 \frac{\partial C(u_1,u_2)}{\partial u_1} = 0 \tag{4}$$

$$\frac{\partial \Pi_j}{\partial u_{i2}} + \sum_{k=1}^{m} \frac{\partial \Pi_k}{\partial u_2} - \lambda_1 \frac{\partial C(u_1,u_2)}{\partial u_2} = 0 \tag{5}$$

After elimination of the Lagrange multipliers, the first-order conditions reduce to

$$\frac{\partial\, C(u_1, u_2)/\partial v_2}{\partial\, C(u_1, u_2)/\partial v_1} = \frac{\partial \Pi_i/\partial u_{i2}}{\partial \Pi_i/\partial u_{i1}} + \sum_{k=1}^{m} \frac{\partial \Pi_k/\partial u_2}{\partial \Pi_k/\partial u_{k1}} \ ; \ i = 1, \ldots, m \tag{6}$$

The revenue function of the firm $\mathcal{F}(\Pi_1,...,\Pi_m)$ may be defined in terms of individual revenues. Fortunately, the first order conditions result in multipliers for the derivatives of Π_i of $\partial\mathcal{F}(\Pi_1^*,...,\Pi_m^*)/\partial\Pi_i$, and equation (6) remains unchanged. Thus even with the realistic assumption of a nonlinear corporate revenue function, the first order conditions become linear additive. Thus the organization's performance $\mathcal{F}(\Pi_1^*,...,\Pi_m^*)$ is maximized for the optimal solution to its user's problem — i.e. objective (1) is reconciled with objective (2). Since the vehicle by which usage information is disseminated is a Walrasian mechanism consisting of allocated costs, this solution is also informationally efficient, and satisfies objective (3).

If there were never a "bottleneck" in accessing equipment with limited capacity, then marginal rates of substitution for the resources would equal the relative costs which the individual users *would be willing to pay* for the Pareto optimal (u^*_{i1}, u^*_{i2}) amounts of processing *if they knew costs* $C(u_1, u_2)$, inducing a schedule of allocated cost d_1 and d_2 such that

$$\frac{\partial \Pi_i/\partial u_2}{\partial \Pi_i/\partial u_1} = \frac{d_2}{d_1} \ ; \quad i = 1, \ldots, m \tag{7}$$

i.e. the ratio of marginal rates of substitution between raw and processed material equals the ratio of allocated costs. These are the equilibrium allocated costs which are ascertained if users are unaware that they are sharing the equipment; i.e. if objective (3) is assumed to be met absolutely.

But in general, users do know about each other through congestion of the equipment, and if $(u_1^*=\sum u_{i1}^*,\ u_2^*=\sum u^*_{i2})$ is to be the Pareto-optimal usage level for the organization, then the relative allocated costs actually charged by management for production level (u^*_1,u^*_2) must be

$$\frac{d_2}{d_1}(1-\delta) \tag{8}$$

rather than d_2/d_1; where the added cost of waiting to use the equipment is

$$\delta=-\frac{d_1}{d_2}\sum_{k=1}^{m}\frac{\partial\Pi_k/\partial u_2}{\partial\Pi_k/\partial u_{k1}} \tag{9}$$

Since $\partial\Pi_k/\partial u_2$ is negative, δ is positive. The optimal allocated cost d_2 that the user is willing to pay essentially has two parts; $d_2(1-\delta)$ is the Pareto optimal allocated cost which should be charged by management for use of the equipment; and δd_2 is the user's efficiency loss due to waiting for the machine; i.e. it is the nuisance cost due to congestion.

Compare this to the standard cost allocation approach. Let $C(u_1,u_2)=\Delta-\alpha_1u_1-\alpha_2u_2=0$ where Δ is the budget for using the equipment. Then the costs allocated to user i are based upon some cost driver ϵ_i, and are α_1/ϵ_i and α_2/ϵ_i. These costs completely ignore: (1) the revenue functions of the users, and thus their motivation for consuming at various levels u_1 and u_2; and (2) the fact that the equipment congests, and that congestion is itself an implied nuisance cost. Where C(.,.) is non-linear, even further situation specific complexities arise. Thus, in traditional cost allocations, equipment usage is virtually assured of being suboptimal.

3: A Linear Approximation for Optimal Allocations

This section provides an approach that approximately satisfies the three stated objectives, and at the same time allows the calculation of unit allocated costs (i.e. is linear). Assume that linear allocated costs c_i (as opposed to "ideal" allocated costs d_i from the prior section) are set by management for equipment resources. The potential users of the equipment are allowed a total discretionary budget Δ per period for the use of the equipment resources. This budget may be subdivided on an individual user or functional department basis; this research makes the more general assumption of a collective budget for the users as a whole. Management updates unit allocated costs c_i for each resource on a periodic basis. Together, Δ and c_i define an allocated cost schedule.

Users receive information about the desirability of using a particular piece of equipment through two sources — (1) feedback about service quality through use, e.g. through the nuisance cost of congestion, and (2) the effect of equipment resource costs and budgets reported to them by the managerial accounting or financial reporting systems. The users are often unaware of out-of-pocket costs $C(u_1,u_2)=0$ because these are never reported to them.

Users replace $C(u_1,u_2)$ in the prior problem with a surrogate linear function of (c_1,c_2,Δ); i.e. the budget variance $\Delta-\sum\sum c_j u_{ij}$; reported by the corporate accounting and performance monitoring systems. An equipment cost schedule which is linear in quantity for any given time period is imperative for any implementable valuation and user charge for services. Otherwise, users would spend an inordinate amount of time jockeying for position in the acquisition of services, knowing that perhaps they would be charged more or less depending upon their position in line. The subsequent analysis is valid, though, for differential pricing based on time of usage, for example lower prices during weekends. Whereas $C(u_1,u_2)$ may be nonlinear and complex, equipment usage costs (c_1,c_2) will generally be fixed per unit. Δ is necessarily defined as a part of the cost schedule, since without it, (c_1,c_2) describes only relative costs to the users.

Assume a total budget of Δ for computing services, plus a set of costs (c_1,c_2). Then define the derived revenue $\Xi_i^*(c_1,c_2,\Delta)$ calculated by the cost accounting system under cost structure (c_1,c_2,Δ) such that

$$\Xi_i^*(c_1,c_2,\Delta) = \max_{u_{i1},\, u_{i2}} \Pi_i(u_{i1}, u_{i2}, u_2) \tag{10}$$

$$\textit{subject to} \quad \Delta = \textstyle\sum_{j=1}^{2} \sum_{i=1}^{m} c_j u_{ij} \tag{11}$$

After elimination of the Lagrange multipliers, the first-order conditions reduce to

$$\frac{c_2}{c_1} = \frac{\partial\Pi_i/\partial u_{i2}}{\partial\Pi_i/\partial u_{i1}} + \sum_{k=1}^{m} \frac{\partial\Pi_k/\partial u_2}{\partial\Pi_k/\partial u_{k1}} \quad ; \quad i = 1, \dots ,m \tag{12}$$

and

$$\Delta = \textstyle\sum_{j=1}^{2} \sum_{i=1}^{m} c_j u_{ij}^* \quad ; \quad i = 1, \dots ,m \tag{13}$$

defines the best linear cost structure, where u^*_{ij} are the Pareto optimal usage levels computed in the prior section. Although (c_1,c_2,Δ) provides a improvement over α_1/ϵ_i and α_2/ϵ_i, it is still suboptimal, and the total revenue under $\Delta=\sum\sum c_j u_{ij}$ will always be less than or equal to the Pareto optimal usage level chosen with knowledge of the equipment performance $\{\sum u_1 \leq u_1, \sum u_2 \leq u_2 : C(u_1,u_2)=0\}$.

Where equipment performance, revenue functions and so forth can only be approximated, the allocated cost schedule may also incorrectly indicate to the user that he can consume at resource levels which are not technologically feasible. If the computed revenue Ξ^*_i is not accurately known, then the budget line must be moved further in towards the origin to insure that users will choose only technologically feasible levels of usage. This explains the risk aversion, or conservatism that is often encouraged in setting equipment usage schedules. Of course, the further towards the origin that $\Delta=\sum\sum c_j u_{ij}$ moves, the greater will be the underutilization in the organization's hardware resources with commensurate reduction in user revenue. Thus the organization incurs costs and inefficiencies in providing equipment users information through the linear approximation defined here. On the other hand, this approximation is more efficient and cost effective than traditional cost allocation approaches.

The direction and magnitude of user consumption adjustments to a given cost schedule may be determined as follows. Roy's [1947] Identity provides user i's demand for resource j

$$u_{ij}(c_1,c_2,\Delta) = -\frac{\partial \Xi^*_i(c_1,c_2,\Delta)/\partial c_j}{\partial \Xi^*_i(c_1,c_2,\Delta)/\partial \Delta} \quad ; j = 1,2 \qquad (14)$$

Precise estimates of demand given a particular allocated cost schedule (c_1,c_2,Δ), and revenue function $\Pi(\cdot)$ may be computed with Roy's Identity. It may not be possible, though, to make significant changes in the existing costs and budgets without political costs.

Drastic changes in the allocated costs charged for use of equipment resources tend to incur the ire of users and financial management alike. The allocated cost schedule recommended in this analysis may indeed reflect economic reality; but users will not find sudden changes in unit costs acceptable, and it is doubtful that arguments of Pareto optimality will be understood by many users — particularly if they are antagonized by cost increases. Management would therefore like to make changes in allocated costs in order to move demand incrementally in a Pareto optimal direction. Management can maximize demand change while minimizing changes in the allocated cost schedule via an iterative process of selecting a new cost and budget structure Roy's Identity can be manipulated to compute the magnitude of change in resource usage given a specific change in the allocated cost schedule. $\partial u_{ij}(c_1,c_2,\Delta)/\partial c_k$ gives the marginal change in user i's demand for resource j=1,2 due to a change in the cost of resource k=1,2. Differentiating Roy's Identity, this becomes

$$\Delta u_{ij} \approx \frac{\partial u_{ij}(c_1,c_2,\Delta)}{\partial c_k}\,\Delta c_k = \tag{15}$$

$$\frac{\dfrac{\partial \Xi_i^*(c_1,c_2,\Delta)}{\partial c_j}\cdot\dfrac{\partial^2 \Xi_i^*(c_1,c_2,\Delta)}{\partial \Delta\,\partial c_k} - \dfrac{\partial \Xi_i^*(c_1,c_2,\Delta)}{\partial \Delta}\cdot\dfrac{\partial^2 \Xi_i^*(c_1,c_2,\Delta)}{\partial c_j \partial c_k}}{\left[\dfrac{\partial \Xi_i^*(c_1,c_2,\Delta)}{\partial \Delta}\right]^2}\cdot\Delta c_k$$

$$= \frac{u_{ij}\cdot\dfrac{\partial^2 \Xi_i^*(c_1,c_2,\Delta)}{\partial \Delta\,\partial c_k} - \dfrac{\partial^2 \Xi_i^*(c_1,c_2,\Delta)}{\partial c_j \partial c_k}}{\dfrac{\partial \Xi_i^*(c_1,c_2,\Delta)}{\partial \Delta}}\cdot\Delta c_k$$

where derivatives of equipment user's revenue reported by the cost accounting system Ξ_i^* are calculated for the current cost structure. Assuming minimal regularity conditions on Ξ_i^*, this last formula is another way of writing the Slutsky Equation in differential format (Cook [1972], Varian [1984]). The leftmost term in the difference describes the change in income from procuring more or less of resource j; the rightmost term describes the substitution of resources j and k. Where the equipment user's revenue function is not well specified, equation (15) supports an iterative approach to adjusting demand. The budget line $\Delta=\sum\sum c_j u_{ij}$ in figure 1 can be moved away from the origin, the resulting demand evaluated for technological feasibility and desirability, and then further adjustments may be made to the resource costs.

4: An Approximation for General Application

This section presents an approximation of typically encountered real world situations, and develops the associated cost allocation structure for this approximation. Although the setup assumed here may vary slightly from reality, the allocated costs computed below will be an improvement over those from existing cost allocation schemes.

Assume equipment utilization that conforms to a single line, single server Poisson process; this may often provide an acceptable approximation to actual machine usage. Define the stated equipment capacity ω and proportion of total capacity used $\rho = u_2/\omega$. In the short run, ω is fixed, the average queue size is $\rho^2/(1-\rho)$ and the probability that the server is free is $1-\rho$ (Lindley [1965]). The average delay may be assumed to be proportional to the average queue size exclusive of the user. Assume, for clarity of example, that the *m* users are homogeneous in preferences and usage; this provides a reasonable approximation of reality given that common service degradation costs δc_2 cause individual differences to be less pronounced. Define *m* identical user revenue functions which include u_{i1}, u_{i2} and u_2

$$\Pi_i(u_{i1}, u_{i2}, u_2) = \beta_1 u_{i1} + \beta_2 u_{i2} - \beta_3 \frac{(u_2/\omega)^2}{1-u_2/\omega} \qquad (16)$$

where $\beta_1, \beta_2, \beta_3 \geq 0$ are raw material, processing and delay costs per unit; delay cost is an approximate measure of the opportunity cost of having the user wait form the machine rather than doing something else productive.

Let $C(u_1, u_2) = \Delta - \alpha_1 u_1 - \alpha_2 u_2 = 0$ where Δ is the budget. Then the costs allocated to user i are based upon some cost driver ϵ_i, and are α_1/ϵ_i and α_2/ϵ_i. First order conditions given the revenue function and cost function are

$$\frac{\partial\, C(u_1, u_2)/\partial y_2}{\partial\, C(u_1, u_2)/\partial y_1} = \frac{\beta_2}{\beta_1} + \frac{m\beta_3}{\beta_1 \omega}\left[1 - \frac{1}{(1-u_2/\omega)^2}\right] \qquad (17)$$

where $u_2 \leq \omega$ and

$$\delta = -\frac{d_1}{d_2} \sum_{j=1}^{m} \frac{\partial \Pi_j/\partial u_2}{\partial \Pi_j/\partial u_{j1}} = \qquad (18)$$

$$= -\frac{m\beta_3}{\beta_2 \omega}\left[1 - \frac{1}{(1-u_2/\omega)^2}\right] \quad ; u_2 \leq \omega$$

Note that in this example, the usage delay term δ is increased by adding users, i.e. increasing *m*, and by making delay more costly, i.e. by increasing β_3, the "delay cost" to the users production. The term δ increases rapidly as usage approaches capacity; where $\delta \geq 1$, even free access to the equipment will not entice additional usage of the resource.

Both d_1 and d_2 are nonlinear in materials usage, and thus, for practical implementation, it is necessary to derive linear approximations $c_1 \approx d_1$ and $c_2 \approx d_2$ which may actually be used as allocated costs in practice through resource costs and a budget (c_1, c_2, Δ) given by management. This may be done by setting $c_1 \approx d_1(u_1^*, u_2^*)$ and $c_2 \approx d_2(u_1^*, u_2^*)$ with optimal production levels (u_1^*, u_2^*).

The approximate allocated costs c_1 and c_2 behave as follows. In the absence of "bottlenecks," i.e. congestion delays, the allocated costs c_2 and c_1 required to achieve optimal production (u_1^*, u_2^*) should be in ratio β_2/β_1; the allocated costs should just use up the budget, so $\Delta - \gamma u_1^* - \gamma(\beta_1/\beta_2)u_2^* = 0$ for some γ. Then $\gamma = \Delta/(u_1^* + \beta_1/\beta_2 u_2^*)$ the allocated cost charged for raw material (u_1) is $\Delta/(u_1^* + \beta_1/\beta_2 u_2^*) = c_1$ and the allocated cost charged for processed material (u_2) is $\beta_1\Delta/(\beta_2(u_1^* + \beta_1/\beta_2 u_2^*)) = c_2$. But where is non zero delay, or where $\beta_3 > 0$, then the allocated cost charged for raw material (u_1) is c_1 and the allocated cost charged for processed material (u_2) is $c_2(1-\delta)$; $c_2(1-\delta)$ is the approximately optimal allocated cost which should be charged by management for use of the equipment; and δc_2 is the implied charge (i.e. not actually collected) from user's opportunity cost of waiting for the machine. Given the first order conditions analogous to those used to compute d_2/d_1

$$\frac{c_2}{c_1} = \frac{\beta_2}{\beta_1} + \frac{m\beta_3}{\beta_1\omega}\left[1 - \frac{1}{(1-u_2/\omega)^2}\right] \quad ; u_2 \in (0,\omega) \qquad (19)$$

Define

$$R = \frac{\dfrac{\beta_2}{\beta_1} + \dfrac{\beta_3 m}{\beta_1\omega} - \dfrac{c_2}{c_1}}{\dfrac{\beta_3 m}{\beta_1\omega}}$$

Then solving equation (19) for u_2 gives

$$u_2(c_1, c_2, \Delta) =$$

$$\omega\left[1 - \sqrt{\frac{\dfrac{\beta_3 m}{\beta_1\omega}}{\dfrac{\beta_2}{\beta_1} + \dfrac{\beta_3 m}{\beta_1\omega} - \dfrac{c_2}{c_1}}}\right] = \omega\left[1 - R^{-1/2}\right] \qquad (20)$$

$u_2(c_1,c_2,\Delta) < u_2^*$ is the total actual production that the users will choose given allocated cost structure (c_1,c_2,Δ); idle capacity will be $u_2^* - u_2(c_1,c_2,\Delta)$. Assume that the budget Δ is large enough to buy all of the available resource 2; thus $\Delta \geq c_2\omega$ and

$$\Delta = \sum_{j=1}^{2} \sum_{i=1}^{m} c_j u_{ij} = \tag{21}$$

$$= c_1 \sum_{i=1}^{m} u_{i1} + c_2\, \omega\left[1 - R^{-1/2}\right]$$

gives

$$\sum_{i=1}^{m} u_{i1}(c_1,c_2,\Delta) = \frac{\Delta}{c_1} - \frac{c_2}{c_1}\, \omega\left[1 - R^{-1/2}\right] \tag{22}$$

The assumption of homogeneous users gives

$$u_{i1}(c_1,c_2,\Delta) = \frac{\sum_{i=1}^{m} u_{i1}(c_1,c_2,\Delta)}{m} \tag{23}$$

and

$$u_{i2}(c_1,c_2,\Delta) = \frac{u_2(c_1,c_2,\Delta)}{m} \tag{24}$$

Substitute these values into the cost accounting systems revenue calculation to derive

$$\Xi_i^*(c_1,c_2,\Delta) = \tag{25}$$

$$= \frac{\beta_1 \Delta}{m c_1} - \frac{\beta_1 c_2}{m c_1}\omega\left[1 - R^{-1/2}\right]$$

$$+\frac{\beta_2\omega}{m}\left[1-R^{-1/2}\right]-\beta_3\frac{\left[1-R^{-1/2}\right]^2}{R^{-1/2}}$$

If there is no cost of delay; i.e. $\beta_3=0$ or $\omega=\infty$; then user willingness to pay is reflected in any cost structure (c_1,c_2,Δ) such that $c_2/c_1=\beta_2/\beta_1$. Therefore, if management sets $c_2/c_1 < \beta_2/\beta_1$, users will use their budget Δ to consume the smaller of ω or Δ/c_2 units of u_2 and any remaining funds are expended upon raw data u_1; if management sets $c_2/c_1 > \beta_2/\beta_1$, users will use their budget Δ to consume Δ/c_1 units of u_1.

5: Discussion

The prior analysis showed that allocated costs c_1 and $c_2(1-\delta)$ approximately satisfy the three objectives stated at the beginning of the paper, while maintaining a linear, implementable cost structure with well defined characteristics. To reiterate, a cost structure that meets objective (1), *Pareto optimal usage,* is important where a large number of users are concerned, and/or where lines of authority are equivocal, and/or where conflict between users already exists, Pareto optimal usage levels will minimize and ameliorate problems arising from conflicted claims on scarce equipment resources. A cost structure that meets objective (2), *incentive compatibility* is important so that it is to each individual user's benefit to use the equipment to the extent most beneficial to the organization. Incentive compatibility is of fundamental importance in Japanese approaches to setting target costs for product costing and engineering. Hiromoto (1988) describes various approaches currently implemented; the approach derived here for computing target costs, should provide even better control over the usage of scarce equipment resources. A cost structure that meets objective (3), *to minimize intra-user communication*, is important because it is unrealistic to assume that every user has the time or resources to be aware of every other user's schedule.

Although this analysis restricted itself to a simple two good, input/output scenario, all of the formulas presented here may easily be generalized to a vector format, which may address situations with multiple inputs and outputs. Although the formulas are more complex than could easily be implemented in a manual accounting systems, the availability of computers to perform calculations makes the previously developed approach implementable.

Where delays in attaining access to equipment affects users in a complex fashion, the prior analysis provides an effective vehicle for communicating equipment costs and performance. For example, in the last approximation, the delay cost is proportional to expected queue length, resulting in an exceptionally complex relationship of demand to allocated cost. A typical OLS regression on historical demand for equipment access would not have revealed this functional relationship between demand and cost. Thus traditional empirical analysis will generally provide cost values which do not reflect equipment performance, and this in turn will result in incorrect analysis from formal and ad hoc decision models which rely on empirically derived costs.

The research showed that the application of traditional cost allocation formulas will in practice tend to result in decisions which underutilize scarce equipment resources. Even though traditional cost allocation schemes may show that particular expenditures will be "cost-effective," users still may not be willing to pay for these "cost-effective" solutions. These nonlinearities may invalidate many of the results derived from otherwise correctly specified cost models produced through traditional cost allocation approaches. Through the calculations of allocated costs c_1 and $c_2(1-\delta)$ provided above, this research adds to the cost analysis discipline an important tool for controlling the usage of scarce equipment resources.

6: Bibliography

Cook, P. (1972) A One Line Proof of the Slutsky Equation, American Economic Review, v. 42, p. 139

Hage J., A. Aiken and C.B.Marrett (1971) Organization Structure and Communications, American Sociological Review, v. 36, 108-129

Hiromoto, T. (1988) Another Hidden Edge – Japanese Management Accounting, Harvard Business Review, July-August

Hurwicz, L (1977) On the dimensional requirements of informationally decentralized Pareto-satisfactory processes, in K.J.Arrow and L.Hurwicz (eds.), Studies in Resource Allocation Processes, Cambridge:Cambridge Univ., pp. 413-24

Jensen, M.C. and W.H. Meckling (1976) Theory of the Firm, Managerial Behavior, Agency Costs and Ownership Structure, Journal of Financial Economics, October, pp. 305-60

Kochen, M. and K.W. Deutsch (1980) Decentralization, Oelgheschlager, Gunn & Hain:Cambridge Mass.

Lindley, D.V. [1965] Introduction to Probability and Statistics From a Bayesian Viewpoint, Cambridge:Cambridge University Press, p.188

Malone, T.W. and Smith, S.A. (1988) Modeling the Performance of Organizational Structures, Operations Research, April-May, 421–36

Osana, H. (1978) On the informational size of message spaces for resource allocation processes, Journal of Economic Theory, v.17, pp. 66-78

Roy, R. (1947) La Distribution de Revenu entre les Divers Biens,

Econometrica, v. 15, pp. 205-225
Samuelson, P.A. (1954) The pure theory of public expenditure, Review of Economics and Statistics, v.36, 387-389
Vancil, R.F. (1978) Decentralization: Managerial Ambiguity by Design, Homewood, IL:Dow Jones-Irwin
Varian, H. (1984) Microeconomics (2nd ed.), New York:Norton
Walker, M. (1977) On the information size of message spaces, Journal of Economic Theory, v.15, pp. 366-75

THE EFFECTS OF DIFFERENT PRODUCTION RATE MEASURES AND COST STRUCTURES ON RATE ADJUSTMENT MODELS

Dan C. Boger and Shu S. Liao*
Naval Postgraduate School

ABSTRACT

The effect of production rate on the cost of weapon systems has attracted much attention in the cost estimating community in recent years. A variety of adjustments to weapon systems cost models have been proposed to reflect the impact of different production rates. The most popular solution is to add a rate term to the traditional learning curve model. This paper examines the effects of different rate measures and cost structures on rate adjustment models. Numerical examples illustrate that the production rate term should be measured as a ratio and not as an absolute quantity of a production lot or a period. The paper also points out that a rate adjustment model is appropriate only with data collected from plants which have not undergone changes in cost structure.

INTRODUCTION

The effect of production rate on the cost of weapon systems has attracted much attention in the cost estimating community in recent years. A variety of adjustments to weapon systems cost models have been proposed to reflect the impact of different production rates. The most popular solution is to add a rate term to the traditional learning curve model. The resulting learning curve model augmented with the production rate variable is usually referred to as a rate adjustment model. The purpose of this paper is to examine the theoretical underpinning of the production rate effect on weapon system

* This research was sponsored by The Cost Analysis Division, Naval Air Systems Command and the Naval Postgraduate School.

cost and illustrate that the popular solution to the rate problem may result in erroneous conclusions. Numerical examples will be used to illustrate the potential problems of the popular approach to production rate adjustment. The paper concludes with a discussion of the scenarios in which the rate adjustment models may be utilized.

CONCEPTUAL FOUNDATION OF PRODUCTION RATE EFFECT

The conceptual foundation of the production rate impact on cost is related to economies of scale. In many industries that effect is well understood. High production rates allow greater use of facilities and greater specialization of labor. The increased volume of materials purchased reduces their unit cost. The increased volume of production activities spreads fixed overhead costs over a larger quantity of products produced. Taken together, all these effects work to increase efficiency and lower production costs (Bemis, 1981; Large, et al., 1974; Linder and Wilbourn, 1973).

It should be noted, however, that a plant with a higher production rate does not necessarily produce at a lower unit cost when compared to another plant. This point is illustrated in Figure 1. Assume there are three plants capable of producing the same item, such as a missile. The Average Unit Cost curve for each plant is shown as AUC1, AUC2, and AUC3, respectively. If the output quantity were fixed at 25 units, then Plant 1 is the most efficient of the three plants. However, if the output level were fixed at the rate of 40 units per period, Plant 1's unit cost would be higher than that of Plant 2, which is the most efficient of the three at that production quantity. This is consistent with economic theory, which says that, in general, there are both economies and diseconomies of scale. This phenomenon is recognized by the above analysts and is reflected in their use of this familiar U-shaped average cost curve to incorporate the effect of production rate into weapon systems cost models.

The same theory of economies and diseconomies of scale is applicable to a single plant's expansion when it is operating beyond its efficient capacity level. This scenario has significant

Figure 1
Average Unit Cost and Production Capacity

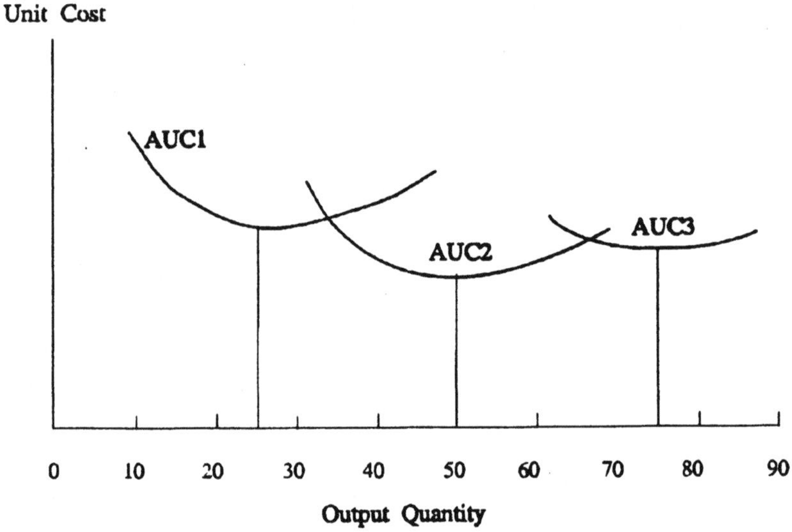

implications in weapon systems cost estimation. Recent experience has shown that production rates of major weapon systems are subject to continual adjustment, sometimes significantly. At the low end of the spectrum is the initial production rate. This is usually a function of early procurement funding constraints and the technical risk of building substantial numbers of newly developed items before the design has fully matured. Thus low rate initial production avoids the risk of incurring costly retrofits to early production units. During this early stage of production, the amount of fixed costs may vary from period to period because of the changing production setup. At the upper boundary is the limitation of available plant capacity and the requirement for additional investments in tooling and facilities for capacity expansion. Additional investments in tooling and facilities alter the cost structure of the plant. The unit cost curve of a plant expanding its investment in tooling and facilities is equivalent to changing from AUC1 to AUC2 as shown in Figure 1.

REVIEW OF RATE ADJUSTMENT MODELS

Although studies of the effect of production rate change on weapon systems cost began as early as the 1950s (Hirsch, 1952; Alchian, 1963), and various models had been proposed, the most widely used rate adjustment model in use today was developed by augmenting the traditional learning curve model with a production rate term:

$$Z = aX^bR^c = YR^c \qquad (1)$$

where,

- Z = unit cost of the item with production rate as well as learning considered,
- X = cumulative quantity produced,
- R = production rate measure,
- Y = unit cost of the item with only learning considered,
- a = a constant, usually called the theoretical first unit cost,
- b = a parameter, usually called the slope of the learning curve,
- c = a parameter, usually called the slope of the production rate curve.

Empirical work on this production rate/learning model was first conducted at RAND, but the model was later popularized by Bemis (1981). Large, et al. (1974) attempted to develop this model for various production cost elements. They were forced to conclude, however, that the production-rate/cost relationship could not be predicted with any reasonable degree of confidence. For production planning purposes, they recommended that production rate effects in aircraft production programs be ignored because they were dominated by other effects. They also suggested that production rate is subject to change and, hence, is difficult to predict.

Further work on the production rate/learning model was carried out by Smith (1976). He analyzed three aircraft programs for which a large number of data values were available due to long production periods. Where the data permitted, Smith applied his model separately to fabrication and assembly labor hours. He then compared his production rate/learning model to a reduced, learning-only model. Smith found that the rate term was an important contributor to the explanatory power of the model. However, he obtained a surprisingly large variation in parameter values for cases with similar production quantities and rates. Additional efforts using this approach were carried out by Bemis (1981), Cox and Gansler (1981), and others.

If one recognizes the inherent rate instability scenario of major weapon systems production and the resultant changing cost structure discussed in the preceding section, then none of the inconclusive findings discussed above would be surprising. In the following sections, we will examine the issues of alternative production rate measures and changing cost structures, and we will discuss other major considerations that must be addressed before one can use the rate adjustment model in weapon systems cost estimation.

ALTERNATIVE PRODUCTION RATE MEASURES

Although the concept of production rate is clear, its measurement is by no means unambiguous. Several alternatives have been used as surrogate measures of production rate. The two primary measures are lot size and annual/monthly production quantity. We will first discuss these two and related measures, along with the difficulties of their use. We then discuss a third alternative, a ratio measure which we believe will avoid some of the difficulties of the measures used to date.

Using Lot Size or Annual/Monthly Quantity As the Rate Measure

Hirsch (1952), Cox and Gansler (1981), and Bohn and Kratz (1984) all used lot size as their measure of production rate. Hirsch was careful to note that his lot intervals were fairly stable; however, this has not been the case with almost all more-recent aircraft programs. Since the time (and, hence, cost) required to produce sequential, similarly-sized lots often changes over the life of the program, it is unclear what is being measured by the lot size proxy.

Perhaps the most common measure of rate is that of production quantity in some time interval. The time period involved is usually selected as a function of data availability. Most studies use annual quantities as a measure of production rates. An inverse of the quantity-per-unit-time measure has also been used; Large, et al. (1974) used the number of months required to reach a certain cumulative production quantity as their inverse measure. Some studies, such as Womer (1984), use monthly data. Womer notes that if there is substantial work-in-progress and the production period is long compared to the period of observation, then units produced in the following time period actually reflect work performed in the preceding time period, and this can result in substantial bias in estimation. Since this problem is especially critical for monthly data, Womer used a lagged model of production to obtain his estimates.

When analyzing a cross-section of programs, it is possible to use an average rate for each program. Because the production rate may change in a typical production run, an average rate for an individual program is usually used in these cross-section analyses. Use of an average may understate the effects of these disruptive rate changes, but we do not expect it to mask the effect of production rate itself. Large, et al. (1974) used this approach in their examination of several programs.

Gulledge and Womer (1986) noted that cumulative quantity is highly correlated with any of the production rate measures discussed above. Hence, using either the lot size or monthly/annual quantities as the measure of R in Equation (1) will produce unreliable models due to this collinearity of the cumulative quantity measure of learning (X) and the measure of production rate (R). The presence of this collinearity has resulted in the inability of analysts to separate statistically the effects of learning and production rate. For example, Large et al. (1974) concluded that the influence of production rate could not be estimated with confidence.

Using a Ratio as the Rate Measure

An alternative to the above measures which will tend to

mitigate the multicollinearity problem is that of a ratio of the above production rate measures. This use of a ratio, if keyed to a base production rate, as the rate curve measure appears to be an innovation in the literature. Bemis (1981) uses the ratio of new rate to present rate as the rate measure, which is more a measure of rate change than a measure of the rate per se. A similar measure was adopted by Balut (1981) and Balut, et al. (1989); they used a ratio of old-to-new lot sizes to account for rate effects in an aircraft repricing model which also included a learning curve. On the other hand, Boger and Liao (1988) proposed using a standard, base, or predetermined rate as the denominator in the ratio and either lot sizes or annual/monthly quantities as the numerator. The advantage of using a base rate is that if one uses the rate to which the manufacturer has tooled the production facility as the base rate, then ratios greater than unity would indicate decreasing returns to variable inputs and ratios lesser than unity would indicate increasing returns to variable inputs.

In addition to the mitigation of statistical problems, the use of a ratio as the rate measure has some intuitive advantages for cost estimating purposes. While the general formulation shown in Equation (1) for production rate is widely used, little has been done to examine the empirical implications of adding the production rate factor to the well known learning curve model. The definition of the parameter **a** of Equation (1) (referred to as the theoretical first unit cost in learning curve theory) is the unit cost when $X=1$ and $R=1$. While this interpretation seems logical, it does result in some awkward numbers because $R=1$ is not close to the relevant production range for most of the production rate measures used in practice. It is, however, for our proposed measure. This issue can be illustrated with a simple example. This example will use a minimum of data points since this is the typical situation faced by cost analysts.

An Illustrative Example

Assume that the data for the first two production contracts for a new weapon system are as follows:

Lot #	Quantity	Unit Price	Algebraic Lot Midpoint
1	100	$43,773	33.9
2	100	31,035	147.0

The algebraic lot midpoint is that quantity on the learning curve which corresponds to the average cost for that entire lot. Liao (1988 and 1989) provides detailed discussions of this concept and its measurement.

A. Ratio Rate Measure -- Since there are only two data points, only the learning curve slope may be estimated at this point. We may use the following formula to determine the learning curve slope:

$$b = \frac{\text{Log } (Y_2 / Y_1)}{\text{Log } (M_2 / M_1)} \qquad (2)$$

where Y_i and M_i represent the unit price and the algebraic midpoint of each lot respectively. The slope of the learning curve for our illustrative data may now be determined as follows:

$$b = \frac{\text{Log } \dfrac{31{,}035}{43{,}773}}{\text{Log } \dfrac{147.0}{33.9}} = -0.234422 \text{ or } 85\% \text{ curve}$$

The first unit cost can be readily obtained by substituting the value of b into the basic learning curve equation:

$$43{,}773 = a\ (33.9)^{-0.234422}$$
$$a = 100{,}000$$

Note that **implicit** in the above computation is the production rate of 100 units. In other words, the $100,000 represents the cost of producing the first unit when the rate is 100 units per year.

Let us assume that for year 3 requirements the government solicits step-ladder quotes from a potential contractor for this system. Step-ladder quotes are the quotes in a schedule of bids from a potential contractor for varying percentages of the government's planned total requirement for that year. (A full set of quotes, using a 10% step, would give the potential contractor's prices for 10%, 20%,. . ., and 100% of the government's requirement.) The differences in the prices quoted by a single contractor for various quantity levels during this single year, in principle, should reflect only the production rate effect. Let us further assume that the slope for the rate curve is 80%. If we want to evaluate the reasonableness of quotes at different production rate levels, the most logical approach is to anchor the rate measure at a given level within the relevant rate range, e.g., 100 units (base rate = 100), and measure different quantity levels as a ratio of that base rate. If the rate curve is known or agreed upon by both parties, the reasonable quotes for various quantity levels may be directly calculated by using the following formula:

$$Z = Yr^d \tag{3}$$

where,

r = the slope of the production rate curve, and
d = the logarithm of R (the ratio measure of rate) divided by the logarithm of 2.

For example, with the assumed 80% rate curve, 85% learning curve, and a=100,000, the reasonable quote for 300 units may be computed as follows:

$$Z_{300} = 25{,}554(0.8)^{\log(3)/\log(2)} = 25{,}554(0.8)^{1.585} = 17{,}942$$

If the parameter value of the rate term is unknown, it can be estimated from annual step ladder quotes as follows. Since we define $Z = aX^bR^c$ or YR^c, the ratio of reasonable bid prices at various quantity levels as a function of the long-term learning curve may be determined as follows:

$$R^c = Z/Y, \text{ or } Z/aX^b \tag{4}$$

We may use the computed ratios for various quantity levels to determine the parameter value for the rate term. Table 1

Table 1

Estimating Rate Effect from Year 3 Step-Ladder Quotes

(a = 100,000, Total Previous Quantity = 200 units)

Quote Quantity	Midpoint	aX^b (Y)	aX^bR^c (Z)	R^c Z/Y	R Q/100
50	224.9	28,088	35,111	1.250	0.5
100	248.4	27,442	27,442	1.000	1
200	293.5	26,390	21,112	0.800	2
300	336.7	25,554	17,942	0.702	3
400	378.6	24,861	15,911	0.640	4
500	419.5	24,269	14,456	0.576	5

shows the procedures described above.

Figure 2 shows the relationship between Z/Y and the rate measure, R. Note that the reasonable quotes should reflect a straight line on a log-log graph as shown in Figure 2. The slope of the rate curve can be derived from the values of the last two columns of Table 1 in the same way that the learning curve slope is usually derived (by using the log-linear regression method). In our case, the regression yields the exponent, **c**, -0.3218, which represents an 80% curve, the slope we used to generate the hypothetical data.

B. Absolute Size Rate Measure -- If we use the lot size or annual/monthly quantity directly as the measure of the production rate, the definition of **a** is necessarily changed to the theoretical first unit cost in the learning curve when X=1 and R=1. Since the rates for the first two buys of our illustrative example are not unity, it is impossible to determine the parameter value of the rate term unless there are at least three, and preferably more, data points.

By combining all available price data when year 3 quotes become available, we can derive the parameter values for the Z equation as shown below:

Figure 2
Production Rate Curve (80%)

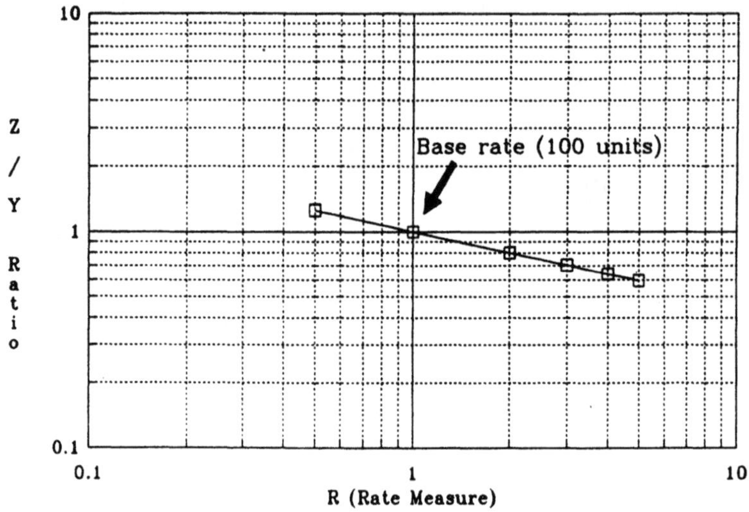

	Ratio Rate Measure	Absolute Size Rate Measure	
a =	\$100,000	\$440,352	
b =	-0.23445	-0.23445	(85% learning curve)
c =	-0.321915	-0.321915	(80% rate curve)

The only difference in results is the first unit cost, **a**. The high value of the first unit cost when using the absolute size rate measure is due to the implicit assumption that it is for X=1 and R=1, which is outside the relevant production rate range and, therefore, is not a meaningful number.

CHANGING COST STRUCTURE

The second major issue facing the use of rate adjustment model for weapon systems cost estimation is the changing cost structure as a result of changes in production setup. Any additional investments in a plant's facilities, whether for capacity expansion or for more efficient production methods, alter the cost structure. This change of cost structure does not create a significant problem for the X term in Equation

(1), since it captures the effect of cumulative production experience (a continuous phenomenon). The changing cost structure, however, poses a serious question about the suitability of using multi-year cost data for cost models involving rate adjustments. The production rate term captures the effect of spreading fixed costs over varying numbers of units. During the early stages of production, the amount of fixed costs may vary from period to period because of the changing production setup. Therefore, the effect of production rate on unit costs may not stabilize until after the production setup and its inherent cost structure is stabilized. Trying to derive a rate curve with historical data from only the early stages of production is probably unreliable.

Figure 3
Total Cost vs Production Rate

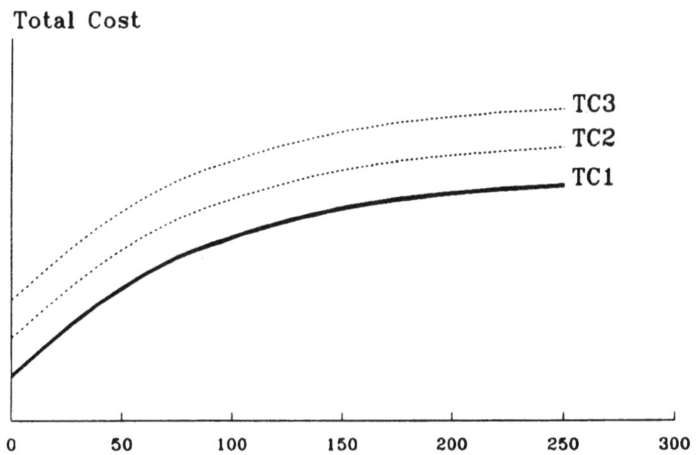

Let us extend the previous example by assuming that the plant capacity is expanded in year 3 to accommodate the higher quantity required. The resultant higher fixed costs push up the total production curve for any given quantity level from TC1 to TC2, as shown in Figure 3. TC3 represents the total cost curve if the capacity is further expanded. Figure 4 depicts the cost reduction curves under different production rates after the learning curve effect has been considered (see Column 3 in Table 1).

Figure 4
Changing Cost Structure & Rate Curve

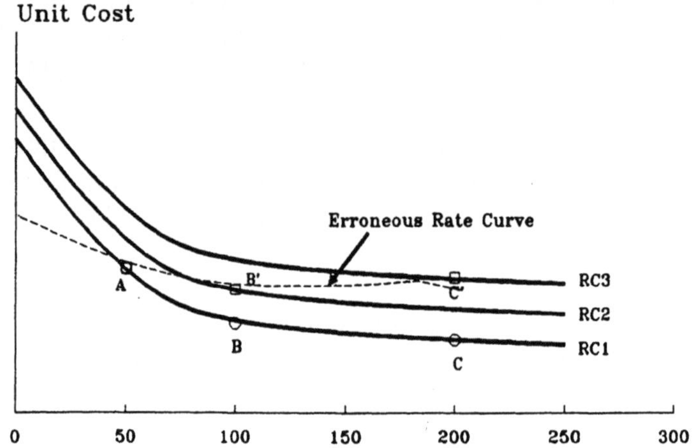

If the government procured 50 units in year 1 under the cost structure labeled TC1 and RC1, 100 units in year 2 under TC2 and RC2, and 200 units in year 3 under TC3 and RC3, the unit costs to the government, after considering the learning curve effect, would be Points A, B', and C'. Deriving a rate curve using A, B', and C' would result in an erroneous rate curve, as shown in Figure 4. The slope of the erroneous rate curve is biased by the changing cost structure.

On the other hand, if there is no change in the plant's cost structure, the same cost curve (TC1 or RC1) applies to years 1 through 3, and the three data points (A, B, and C in Figure 4) would all fall on the same curve (RC1). Therefore, the data would be appropriate for estimating the parameter value for the rate term. The same is also true for step-ladder quotes for any particular year, which reflect the spreading of fixed costs in a particular year (Points A, B, and C) and, therefore, are also appropriate for estimating the parameter values using Eq. (1).

Table 2

The Effect of Changing Cost Structure on Unit Costs
(a = 100,000, LC = 85%, R = 80%)

Lot #	Quantity	Total Cost	Unit Cost
A. Same Cost Structure:			
1	50	$3,191,511	$63,830
2	100	3,437,523	34,375
3	200	5,601,184	18,671
B. Changing Cost Structure:			
1	50	$3,191,511	$63,830
2	100	3,837,523	38,375
3	200	6,401,184	21,337

The issue discussed above can be illustrated with a numerical example as shown in Table 2. Data for Scenario A are constructed by assuming that there was no change in the cost structure in the contractor's plant. Data for Scenario B are constructed by adding $400,000 and $800,000 of additional fixed costs to year 2 and year 3 total costs respectively. Using the three data points under each scenario to derive the parameters for Equation (1) results in the following:

	Scenario A	Scenario B:
a =	$100,000	$72,227
b =	-0.2344 (85%)	-0.1389 (91%)
c =	-0.3219 (80%)	-0.3959 (76%)

It can be seen clearly that analysis of data from Scenario A results in correct parameters, while analysis of data from Scenario B distorts all three parameters. What we can conclude is that using cost data obtained from a plant which has experienced a changing cost structure violates the statistical requirement of drawing samples from a homogeneous population. The consequence of sampling from different populations is the distortion of all parameters, as shown above.

CONCLUSIONS

In this paper, we examined the conceptual underpinning of the production rate effect on weapon system costs as well as various production rate measures for rate adjustment models. The first conclusion is that the production rate term should be measured as a ratio, not as an absolute quantity of lot size or annual/monthly quantity. Expressing the production rate as a function of a base rate within the relevant range allows the analyst to estimate the learning curve from scanty historical data with more confidence as well as adjust costs for the applicable rate effect. It also facilitates the comparison of current step-ladder quotes with the historical contract awards.

There are several other practical considerations that favor the use of a ratio as the rate measure. The data base available for learning curve and rate curve determination is typically scanty. Using unity as the rate base requires both X and R as the independent variables in parameter determination. Having to use two independent variables reduces the degrees of freedom and increases the estimating error accordingly.

The second conclusion is that a stringent condition must be met before an analyst can use multi-year cost data to derive parameter values for the widely used rate adjustment model (Equation 1). The condition is that the underlying cost structure (variable/fixed cost mix and direct/indirect cost mix) must remain the same for all time periods covered by the data. This condition is met by step-ladder quotes for various quantities within the same period or by a plant that has stabilized its production capacity and setup. Unless this condition is fulfilled, the rate adjustment cost model may significantly distort the parameters. We believe that the inconclusive findings of prior research regarding production rate impact on weapon systems cost can be partially attributed to this problem.

REFERENCES

1. Alchian, A. A. (1963), "Reliability of Progress Curves in Airframe Production," *Econometrica*, Vol. 31, pp. 679-693.
2. Balut, S. J. (1981), "Redistributing Fixed Overhead Costs," *Concepts*, Vol. 4, No. 2, pp. 63-72.
3. Balut, S. J., T. R. Gulledge, Jr., and N. K. Womer (1989), "A Method of Repricing Aircraft Procurement," *Operations Research*, Vol. 37, pp. 255-265.
4. Boger, D. C. and S. S. Liao (1988), *Quantity-Split Strategy Under A Competitive Procurement Environment*, NPS-54-88-008, Naval Postgraduate School, Monterey, CA.
5. Bohn, M. and L. A. Kratz (1984), "The Impact of Production Rate on Unit Costs," Paper presented at the ORSA/TIMS Joint National Meeting, San Francisco, CA.
6. Bemis, J. C. (1981), "A Model for Examining the Cost Implications of Production Rate," *Concepts*, Vol. 4, No. 2, pp. 84-94.
7. Cox, L. W. and J. S. Gansler (1981), "Evaluating the Impact of Quantity, Rate, and Competition," *Concepts*, Vol. 4, No. 4, pp. 29-53.
8. Gulledge, T. R., Jr. and N. K. Womer (1986), *The Economics of Made-to-Order Production*, Springer-Verlag, New York, NY.
9. Hirsch, W. Z. (1952), "Manufacturing Progress Functions," *The Review of Economics and Statistics*, Vol. 34, pp. 143-155.
10. Large, J. P., K. Hoffmayer, and F. Kontrovich (1974), *Production Rate and Production Cost*, R-1609-PA&E, The RAND Corporation, Santa Monica, CA.
11. Liao, S. S. (1988), "The Learning Curve: Wright's Model vs. Crawford's Model," *Issues in Accounting Education*, Vol. 3, No. 2, pp. 302-315.
12. Liao, S. S. (1989), "Modifications and Extensions: Applying the Learning Curve Formula," *Program Manager*, Vol. XVIII, No. 3, pp. 40-46, 91.
13. Linder, K. P., and C. R. Wilbourn (1973), "The Effect of Production Rate on Recurring Missile Costs: A Theoretical Model," *Proceedings of the Eighth Annual Department Defense Cost Research Symposium*, Airlie,

VA, pp. 276-300.

14. Smith, L. L. (1976), *An Investigation of Changes in Direct Labor Requirements Resulting From Changes in Airframe Production Rate*, Ph.D. dissertation, University of Oregon, Eugene, OR.
15. Womer, N. K. (1984), "Estimating Learning Curves From Aggregate Monthly Data," *Management Science*, Vol. 30, pp. 982-992.

PRODUCTION LOT SIZING IN A CLASS OF BATCH PROCESS FLOW SHOPS

Avijit Banerjee and Somkiat Eiamkanchanalai
Department of Management & Organizational Sciences
Drexel University
Philadelphia, PA 19104

ABSTRACT

This paper develops a production lot sizing model for a class of batch process flow shops under deterministic conditions. These flow shops are characterized by unit batch transfers between successive work stations, lot-for-lot input item delivery and the possibility of multiple externally procured input items at any work station during the manufacturing process. In formulating the model, the effects of work-in-process inventories resulting from processing at a finite rate at each work center are explicitly taken into account. The concepts developed are illustrated through a simple numerical example.

INTRODUCTION

The advent of new methodologies, such as just-in-time (JIT) and optimized production technology (OPT), for the purpose of production and inventory control has spawned recent research on flow shops in two new directions (Trietsch (1989), Vickson and Alfredsson (1989)). The first of these focus on bottleneck resources and on scheduling activities downstream from bottlenecks, i.e. forward scheduling. The second area addresses the issue of transfer batches, i.e. moving subsets of a total production lot to downstream machines, allowing, to some extent, simultaneous processing of the lot on several machines or work centers, thus reducing the makespan. This paper is related to the latter direction in flow shop research.

In his early work concerning multistage lot sizing, Szendrovits (1975) recognized the advantages of transfer batch sizes that are smaller than the production lot. Subsequent extensions by Goyal (1976), Szendrovits (1976), Drezner et al. (1984) and Truscott(1986) attempt to deal with the questions of optimal manufacturing and transfer batch sizes simultaneously under different assumptions and conditions. More recently, Graves and Kostreva (1986) outline a model for the transfer lot sizing problem within an MRP framework, with the assumption that the production batch quantity is an integer multiple of the former. This restrictive assumption has been removed by Trietsch (1989), who has formulated a generalized technique for the transfer lot sizing problem for a given production lot with a transfer budget limit.

One important rationale for a discrete transfer batch quantity, cited by most authors, is the relatively high cost of moving the units one by one, resulting from the concept of a fixed moving cost (in addition to variable costs), that is independent of the quantity moved. Such a concept often applies in the case of variable path material handling systems. For example, if in-process work is moved by fork lift trucks, a fixed cost is usually allocated to each trip made by such a truck, regardless of the amount of material moved

While such a state of affairs may exist in many traditional manufacturing systems, particularly those outside the realm of mass production, it may not be valid in the case of a number of newer production system configurations that have resulted from recent innovations in technology. In particular, manufacturing systems equipped with fixed path, automated material handling capabilities, such as conveyors, have little or no fixed costs associated with the transfer of in-process work. In the case of flow shops that fit such a description, moving in-process work between work stations one by one, rather than in sub-lots, may indeed be desirable for inventory, as well as makespan reduction.

This paper focuses on the above mentioned class of batch production flow shops. More specifically, we limit our attention to systems that are characterized by an inter-station transfer batch size of unity. It is important to note that although the variety of products manufactured in such systems dictate their processing in substantial batches, the unit batch transfer concept still allows us to capture the major advantages of JIT in terms of reduced work-in-process inventories and lead times. Our major objective in this paper is the development of a manufacturing lot sizing model for an item produced in a make-to-stock batch production flow shop with the above mentioned characteristics, operating under deterministic conditions.

Unlike some of the related work previously done in this area, we allow the possibility of different production rates for the various work stations in the shop. Also, as alluded to above, the phenomenon of unit transfer of the in-process work between work stations implies virtually simultaneous conversion of input to output at finite rates occurring at various stages of production. This is incorporated in our analysis and its impact on the lot sizing decision is ascertained. A third, and perhaps the most important, point of distinction between this paper and previous work in the area is that we allow the possibility of externally procured input, such as materials and components, at any stage of the production process. In contrast, most of the existing flow shop research allow input only at the first stage of production, i.e. the first work station. In this sense, we feel that the results obtained here would have somewhat wider applicability. Finally, it is hoped that some of the concepts developed here will be relevant in the design and control of flexible manufacturing systems (FMS) with the above mentioned characteristics, in view of the increasing importance of such systems in the global manufacturing arena.

ASSUMPTIONS AND NOTATION

The major assumptions made in this paper are as follows:

1. A product is manufactured autonomously in a make-to-stock, batch production flow shop, consisting of multiple machines (work centers) and is stored in a single location.
2. Input items, e.g. materials, parts, etc., may be necessary at several production stages or work centers and all input items are delivered on a lot-for-lot basis, depending on the production lot size.
3. The lot sizes need not be integers.
4. The production and storage environment is deterministic and all the relevant system parameter values are known.
5. As soon as the processing of a unit is completed on a machine, it is immediately moved to the subsequent machine for the next operation, i.e. all transfer batches are of unit size.
6. The production rate at any work center is finite and stationary over time.
7. Different work stations may have different processing rates with respect to the product in question.
8. End product and input item stockouts are not permitted.
9. The demand rate for the product is stationary over time.
10. No quantity purchase discounts are available for the externally procured input items.
11. There are no limitations on storage space and capital invested in inventories.
12. The planning horizon is infinite and the optimization objective is cost minimization.

It should be noted that we make the assumption of lot-for-lot delivery of the input items for the sake of simplicity. Although this limits the applicability of our results, we feel that the insights gained from our analysis, in spite of the simplifying assumptions, will be helpful towards the future development of more realistic models.

The following notation is used throughout this paper:

D = demand rate for the end item (units/period);

Q = production lot size in units;

r = inventory holding cost rate in \$/\$/period;

P_j = production rate at the jth work station (units/period) $j = 1, 2, ..., n$;

$C(o)_j$ = total unit (exclusive of setup) cost of output from work station j (\$/unit);

$C(i)_j$ = total unit cost of all external input (in addition to work-in-progress) required at station j (\$/unit);

S_j = total fixed setup and input ordering cost (if any) incurred at work station j (\$/batch);

$I(o)_j$ = average inventory level of output (units) from work station j;

$I(i)_j$ = average inventory level of external inputs to the jth work station (units);

t_j = production time per cycle at station j in periods;

TRC = total relevant cost in dollars per period.

Note that, of necessity, all $P_j \geq D$ for $j = 1, 2, ...,n$. Without loss in generality, it is assumed that one "unit" of each input item is necessary for producing a unit of the end product. Furthermore, we treat multiple input items at a work station as a single composite item and a "unit" of such a composite item is defined accordingly. For example, if 1 lb. of plaster, 4 bolts and 2 plastic sheets are used as input and incorporated into the product at a given work station, then a "unit" of input at this station consists of these amounts of the above mentioned items. Finally, in order to avoid the complexities associated with joint ordering, we assume that the input items are procured independently.

MODEL DEVELOPMENT

The 2-Station Model

Our analysis follows that of Banerjee and Burton (1989) who deal

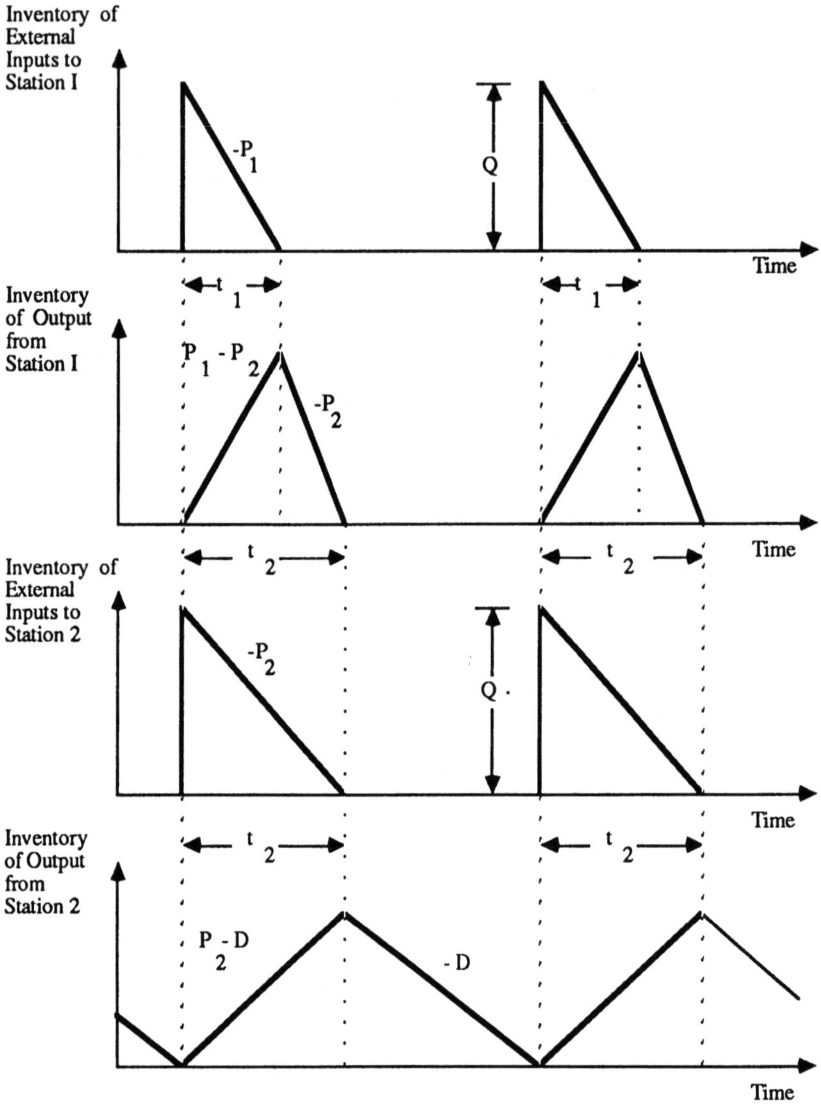

Fig. 1 Inventory time plots for a 2-station flow shop
(case 1: $P_1 \geq P_2$)

with the lot sizing question in single stage production systems. For systematic development, we begin our analysis with a two-station flow shop. The results thus obtained are then easily extended to the case of the more general n-station flow shop.

In the simple 2-station system configuration, two cases are possible on the basis of the relative production rates of the work stations, i.e. $P_1 \geq P_2$ and $P_1 \leq P_2$. The inventory cycles of the various input and output items at the two work centers under these two situations are depicted in Figures 1 and 2, respectively. It is to be noted that in constructing these figures, we assume without loss in generality that the time needed to process a unit of the product at a station and to transport it to the following station is negligibly small.

Case 1: $P_1 \geq P_2$: From Figure 1, the average inventory levels of the input to and the output from the two stations are:

$$I(i)_1 = (Q/2)(t_1)(D/Q),$$
$$I(i)_2 = (Q/2)(t_2)(D/Q),$$
$$I(o)_1 = 1/2\ (P_1\text{-}P_2)(t_1)(t_2)(D/Q),$$
$$I(o)_2 = 1/2\ (P_2\text{-}D)(t_2)(Q/D)(D/Q).$$

Noting that $t_1=Q/P_1$ and $t_2=Q/P_2$, we obtain

$$I(i)_1 = DQ/2P_1$$
$$I(i)_2 = DQ/2P_2,$$
$$I(o)_1 = (DQ/2)(P_1\text{-}P_2)/P_1P_2,$$
$$I(o)_2 = (Q/2)(1 - D/P_2).$$

Case 2: $P_1 \leq P_2$: With respect to Figure 2, analysis along similar lines leads to the following:

$$I(i)_1 = DQ/2P_1,$$
$$I(i)_2 = DQ/2P_2,$$
$$I(o)_1 = (DQ/2)(P_2\text{-}P_1)/P_1P_2,$$
$$I(o)_2 = (Q/2)(1 - D/P_2).$$

The above results may be generalized, regardless of the relative magnitudes of P_1 and P_2, as shown below:

$$I(i)_j = DQ/2P_j, \text{ for } j = 1, 2$$
$$I(o)_1 = (DQ/2)|P_1\text{-}P_2|/P_1P_2,$$

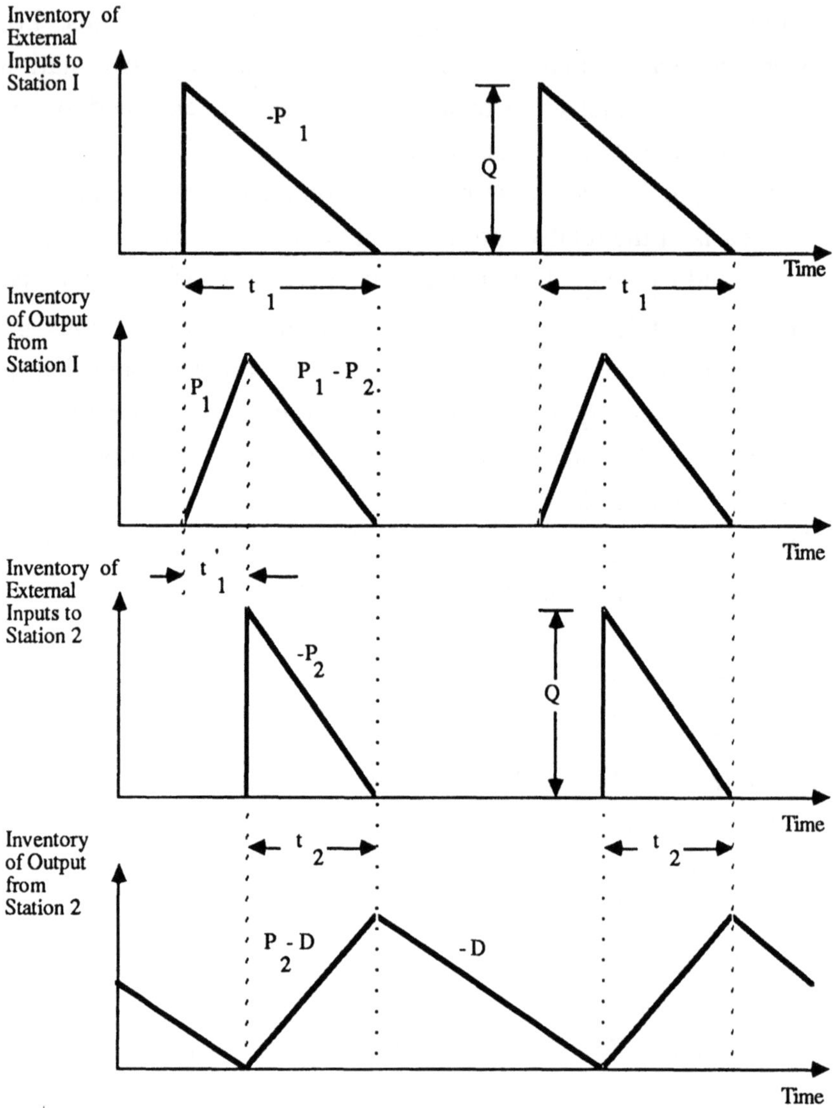

Fig. 2 Inventory time plots for a 2-station flow shop
(case 2: $P_1 \leq P_2$)

$$I(o)_2 = (Q/2)(1 - D/P_2).$$

Using these, the total relevant cost per period resulting from a production batch size of Q in a 2-station flow shop can be expressed as

$$\begin{aligned} TRC(Q) = (D/Q)(S_1+S_2) + rC(o)_1(DQ/2)|P_1-P_2|/P_1P_2 \\ + rC(i)_1(DQ/2P_1) + rC(i)_2(DQ/2P_2) \\ + rC(o)_2(Q/2)(1-D/P_2) \end{aligned} \qquad (1)$$

Note that in (1) the first term represents the total fixed setup and input ordering costs at the two stations, the second and the last terms denote the carrying costs associated with the output from the stations and the remaining terms represent the inventory holding costs resulting from the input items. Cost function (1) can be written as

$$\begin{aligned} TRC(Q) = (D/Q)\sum_{j=1}^{2} S_j + (Qr/2)[C(o)_1D|P_1-P_2|/P_1P_2 \\ + \sum_{j=1}^{2} C(i)_jD/P_j + C(o)_2(1-D/P_2)]. \end{aligned} \qquad (2)$$

The n-Station Model

Extending the cost model (2) above, developed for the 2-station case, to the n-station flow shop (where only a subset K of the n stations require external input items), we obtain

$$\begin{aligned} TRC(Q) = (D/Q)\sum_{j=1}^{n} S_j + (Qr/2)[\sum_{j=1}^{n-1} C(o)_jD|P_j-P_{j+1}|/P_jP_{j+1} \\ + \sum_{k \in K} C(i)_kD/P_k + C(o)_n(1-D/P_n)] \end{aligned} \qquad (3)$$

It can be easily shown that (3) is strictly convex in Q. Thus, in order to minimize TRC, setting the first derivative of (3) with respect to Q equal to 0 at Q = Q* results in

$$Q^* = \sqrt{\frac{2\sum_{j=1}^{n} S_j}{r\left[\sum_{j=1}^{n-1} \frac{C(o)_j |P_j - P_{j+1}|}{P_j P_{j+1}} + \sum_{k \in K} C(i)_k / P_k + C(o)_n (1/D - 1/P_n)\right]}} \qquad (4)$$

Thus, Q* represents the optimal production batch size that minimizes the total relevant cost per period.

Numerical Example

The production batch size model developed above is illustrated through the example of a product that is manufactured in a 3-station flow shop, having the characteristics adopted in this paper. The information presented in Table 1 apply.

The data in Table 1 indicates that two input items are processed in station 1. Station 2 requires no external input and station 3 processes a single input item, in addition to the work-in-process. Thus, the composite input item cost at station 1 is \$20.00/unit (\$15.00 + \$5.00) and the composite ordering cost is \$25/order (\$10 +\$15). Furthermore, the total fixed costs per batch at stations 1, 2 and 3 are, respectively, \$65 (\$40+\$25), \$30 and \$45 (\$37+\$8). Substituting these and the other relevant parameter values in (4), the optimal production lot size is

$$Q^* = \sqrt{\frac{2(65 + 30 + 45)}{.2\left[\frac{30|1000-400|}{1000(400)} + \frac{50|400-500|}{400(500)} + \frac{20}{1000} + \frac{30}{500} + 10\left(\frac{1}{100} - \frac{1}{500}\right)\right]}}$$

= 38.39 units.

In contrast, if the classical production lot size formula is applied to station 3, the final machine in the process yielding the finished product, the lot size indicated is about 22 units. This results from ignoring the effects of input item procurement, intermediate process setups and work-in-process inventories,which are explicitly taken into account in our model, expressed by (4).

TABLE 1
PROBLEM DATA

Stn. j	Production Rate (Units/year) P_j	Production Setup Cost ($)	Output Cost ($/unit) $C(o)_j$	Input Cost ($/unit) $C(i)_j$	Ordering Cost ($/order)
1	1000	40	30	a: 15.00 b: 5.00	10 15
2	400	30	50	-	-
3	500	37	100	3	8

D = 100 units/year, r = $0.20/$/year

SUMMARY AND CONCLUSIONS

In this paper we have developed a production lot sizing model for a class of batch production flow shops. These flow shops are characterized by unit batch transfers and provision for externally procured input at any work center. Some FMS configurations, at least approximately, tend to fit this description. A number of simplifying assumptions adopted in our analysis render our model relatively tractable. However, it can be argued that the same assumptions also limit the usefulness and applicability of the model. Nevertheless, our work is not without some value. We hope that we have been able to shed some light in terms of modeling the behavior of work-in-process inventories and their effects on the batching decision. These insights are likely to be helpful for the development of more elaborate and generalized lot sizing models for batch process flow shops in future research.

REFERENCES

1.Banerjee, A. and Burton, J. S., "Lot Sizing and Work-In-Process Inventories in Single Stage Production Systems", in Gulledge, T. R. and Litteral, L. A. (Eds.), Cost Analysis Applications of Economics and Operations Research, New York: Springer-Verlag, 1989, 283-297.

2.Drezner, Z., Szendrovits, A. Z. and Wesolowski, G. O., "Multistage Production with Variable Lot Sizes and Transportation of Partial Lots", European Journal of Operational Research, Vol 17 (1984), 227-237.

3.Goyal, S. K., "Note On Manufacturing Cycle Time Determination for a Multistage Economic Production Quantity Model", Management Science, Vol. 23 (1976), 332-333.

4.Graves, S. C. and Kostreva, M. M., "Overlapping Operations in Material Requirements Planning", Journal of Operations Management, Vol. 6 (1986), 283-294.

5.Szendrovits, A. Z., "Manufacturing Cycle Time Determination for a Multistage Economic Production Quantity Model", Management Science, Vol. 22 (1975), 298-308.

6.____________, On the Optimality of Sub-Batch Sizes for a

Multistage EPQ Model - A Rejoinder", Management Science, Vol. 23 (1975), 334-338.

7.Trietsch, D., "Polynomial Transfer Lot Sizing Techniques for Batch Processing on Consecutive Machines", Report No.: NPS-54-89-011, 1989, Naval Postgraduate School, Monterey, CA.

8.Truscott, W. G., "Production Scheduling with Capacity Constrained Transportation Activities", Journal of Operations Management, Vol.6 (1986), 333-348.

9.Vickson, R. G. and Alfredsson, B. E., "Two and Three Machine Flow Shop Scheduling Problems with Equal Sized Transfer Batches", Working Paper, 1989, University of Waterloo, Waterloo, Ontario, Canada.

III. Schedule Estimating

SCHEDULE ESTIMATING RELATIONSHIPS FOR AIR-LAUNCHED MISSILES

Bruce Harmon
Institute for Defense Analyses
1801 N. Beauregard Street
Alexandria Virginia

Lisa Ward
General Electric Aerospace
8080 Granger Court
Springfield, Virginia

BACKGROUND

Representatives of the Office of the Secretary of Defense (OSD) are responsible for review of service acquisition programs prior to their inclusion in the defense program that is submitted to the president and subsequently to Congress. Part of this process involves the review of proposed acquisition schedules. This paper documents research into methods for assessing the reasonableness of proposed acquisition schedules for air-launched-missile programs. Such methods should reproduce typical schedules while accounting for schedule variations among programs. Variations may be due to a variety of factors, such as the acquisition strategy embodied in the programs and the technical characteristics of the missile.

APPROACH

This work follows on an earlier Institute for Defense Analyses (IDA) study [1] that examined tactical aircraft acquisition schedules. The approach used here in many ways parallels that used for the earlier study. Our approach was to:

- Collect historical schedule and technical data on air-launched-missile programs.
- Present historical missile acquisition program schedules and related data in consistent formats for use in data analyses and for comparison with proposed acquisition programs.
- Perform analyses on schedule intervals in the data, and integrate the analyses into a schedule assessment tool

that spans the period from the start of full scale development (FSD) through early production.

Past research in the area of air-launched-missile schedules has been minimal. More general research in aerospace systems schedules is reviewed in [1].

The examination of historical data is the appropriate starting point for the development of a schedule assessment tool. We collected data on fourteen air-launched-missile programs. Included in our sample are the air-launched-missile programs that involved substantial developments from the mid-1960s to the late 1980s. Our sample contains seven air-to-air and seven air-to-surface systems.

In order to analyze the data, we decomposed the FSD program schedules into four periods for which estimating relationships could be found. (We refer to these periods as "intervals" even though they are not necessarily mutually exclusive; that is, some intervals overlap.) The primary technique in defining and testing these relationships was linear regression analysis. Of relevance were not only schedule data but also the program and technical parameters to which the length of schedule intervals may be related.

The four program intervals we analyzed were: (1) time to first guided launch as measured from FSD start to first guided launch, (2) length of the development flight test program as measured from the first guided launch to the end of initial operational testing, (3) early production time as measured from long-lead and full-funding release for the initial production lots to the first production deliveries for those lots, and (4) program length from first launch as measured by the time from first guided launch to first production delivery. Figure 1 shows the relationship of these four intervals for an example program, the AIM-120A. Also shown are more detailed milestones.

An issue that arose when determining the intervals to be analyzed was what milestone should mark the end of development. Because definitions of initial operational capability (IOC) differ among programs, and because some inconsistencies in the relation-ship between IOC dates and other program milestones were unexplained, we chose not to use IOC to mark development completion. Another possibility for development end was the completion date of the guided-launch test program through initial operational testing. If we

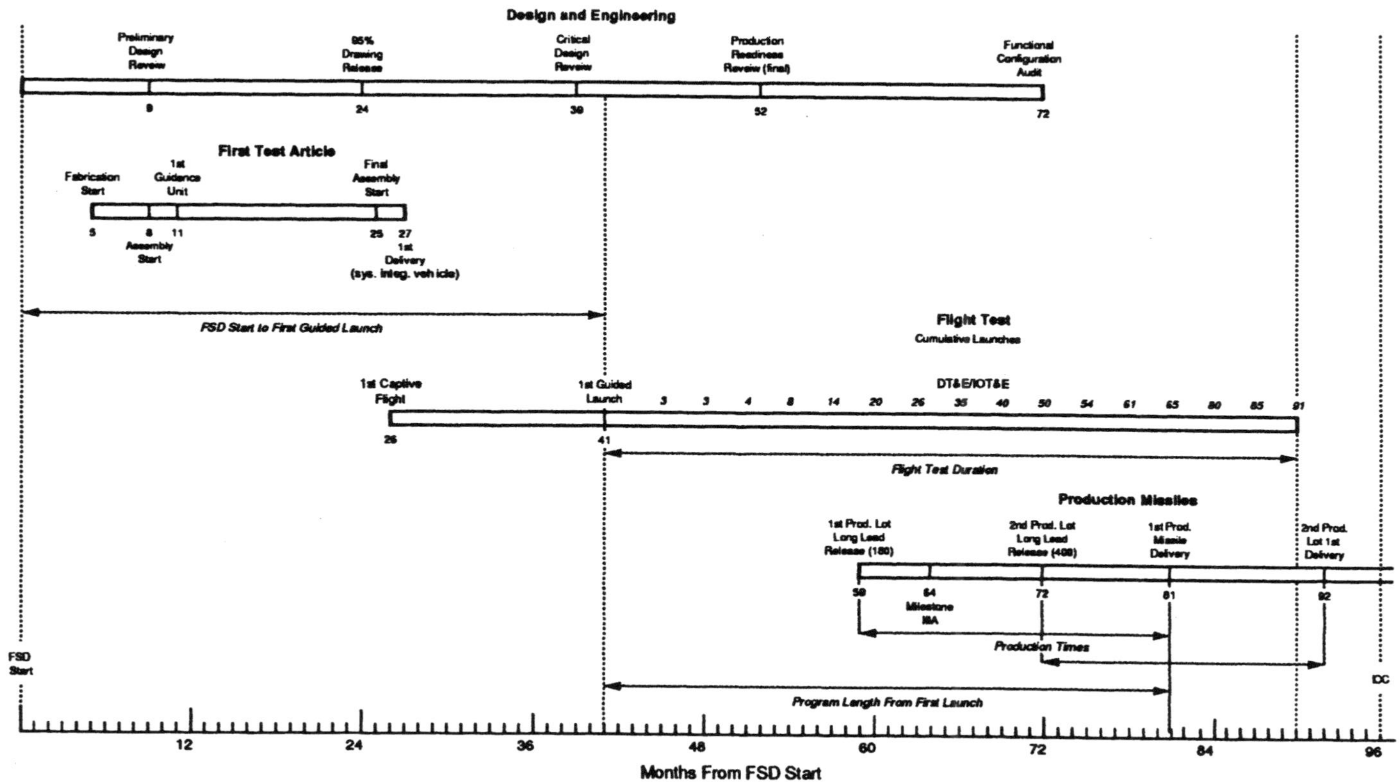

Figure 1. AIM-120 Schedule

considered this the completion of development, an estimate of total development length could be made by simply adding the estimated length for time to first guided launch to that for the flight test program. However, we were searching for a milestone related to the availability of missiles for operational inventories. Because production start, and hence the delivery of production missiles, is not tied to a test milestone common to all programs, test program end was not a consistent indicator of the availability of operational missiles. Production milestones are related to the test program through program concurrency, i.e., the overlap of the test program with initial production activity. The degree of overlap (which we refer to as "concurrency") can vary widely between programs and, within limits, is determined by policymakers.

In the earlier study on tactical aircraft schedules, we used the delivery date of a quantity of aircraft associated with squadron size, 24, as the development end date. Unfortunately, the inventory requirements and the production rates associated with different types of missiles vary widely, so using a milestone associated with a fixed number of production deliveries would lead to inconsistencies across programs.

For those reasons, we decided to use the delivery date of the first production missile to mark the end of development. Given this definition of development end, the estimating relationships derived from the analysis of schedule interval data can be used to estimate overall development program length in two ways. Both ways use the same equation to estimate time to first guided launch. The preferred way of arriving at the time from first guided launch to first production delivery is to separately estimate flight test program parameters and the production time for the first production lot, and then combine these two estimates (i.e., combine the estimates for intervals 2 and 3) using an explicit measure of concurrency. As an alternative method and an aggregate check on the first method, a single equation for program length from first guided launch (interval 4) was also estimated.

In applying regression analysis, we treated schedule intervals (measured in months) as dependent variables and regressed them against independent variables that were thought to influence these intervals. The adequacy of these regression models was tested using standard measures of statistical significance and model fit. Models whose parameter estimates carried intuitively incorrect signs (for example, if the model indicated decreasing interval lengths with increasing missile weight) were rejected.

The one interval for which length was not directly estimated was the span of the development flight test program. Here the dependent variable was the average monthly launch rate by test program phase. Given this value and information about the number and employment of test missiles, the length of the flight test program could be determined.

DATA COLLECTION

Data were collected in some detail for fourteen air-launched-missile programs. Criteria for selecting a program for inclusion were the newness of the program, its importance in historical perspective, and the expected availability of data. All air-launched-missile programs involving major developments that occurred from the late 1960s to the late 1980s are included in the data base. These programs include both air-to-air and air-to-surface systems.

In our data collection effort, emphasis was placed upon the full scale development (FSD) phase of the acquisition cycle. Our primary concern was with FSD program milestones and the schedule intervals derived from them. Schedule intervals in the concept exploration phase and the demonstration and validation phase prior to FSD are often highly dependent upon political factors and were therefore not emphasized in our data collection or analyses. However, pre-FSD prototype developments seem to have had consistent patterns across programs and because such development will be important in the acquisition strategies for future systems, they are therefore included. We defined prototype programs as pre-FSD development efforts where test missiles were air-launched. Some emphasis was placed on collecting production data, particularly data relevant to the initial production build-up.

Often ambiguities arise when defining program phases with which program milestones are associated. Air Force programs tend to follow the classic acquisition pattern with clear delineations between the advanced development/prototype phase (Milestone I to Milestone II), the FSD phase (Milestone II to development end), and initial production. In Navy programs there is often not a clear break between what we call the "prototype phase" and the FSD phase. Also in Navy programs, the delineation between FSD and production is unclear. Pilot production contracts are often developmental in nature, providing test missiles and other support for the latter part of the development program.

Statistics characterizing development flight test are important. The length of the flight test program is perhaps the most important determinate of overall development program length. Flight test duration is determined by the number of test missiles launched and the rate at which test launches are accomplished. The most central aspect of a missile's flight testing is the fully guided-launch program. This is where we concentrated our data collection and analyses.

We also collected data on missile physical and performance characteristics to which schedule intervals may be related. We depended on unclassified sources of information wherever possible.

Sources of the data included, the military services, prime contractors, third parties (studies and data bases at IDA, Rand, etc.), and the open literature. Schedule and missile characteristic data were obtained from Selected Acquisition Reports (SARs) and numerous government and secondary data sources [References 2 through 21]. An earlier Rand schedule interval study [2] was an important source of planning phase (pre-FSD) data. A recent IDA study [3] containing program case histories and a thorough compilation of SAR data was also an important source of information. The main sources of schedule, test, and missile characteristic data were the prime contractors and the services. Data requests were sent to the prime contractors asking for detailed missile characteristic, milestone, testing, and production data. Where the response from the prime contractors was poor, we went to the services for additional data. Prominent among service sources was the Naval Weapons Center (NWC).

PROGRAM AND MISSILE CHARACTERISTICS

Table 1 presents information characterizing the seven air-to-air-missile programs in our sample, and Table 2 gives information on their associated missiles. Tables 3 and 4 present corresponding information for the seven air-to-surface programs and missiles. In each case, technical characteristics that might have an effect on schedule intervals are presented. The fourteen programs represent a rich variety in terms of both program and missile attributes. Five different prime contractors are represented. Eight missile programs were preceded by prototype programs. Six of the fourteen programs were modifications of previously developed missiles. Missile physical and performance characteristics vary widely and a variety of guidance types are represented. Although the focus of the study was tactical air-launched missiles, we included two strategic air-to-surface systems, the AGM-69A short-range attack missile (SRAM) and the AGM-86B air-launched cruise missile (ALCM). Both of these systems are distinguishable from the tactical systems in that they have inertial terminal guidance systems. The ALCM is unusual for its very long-range capabilities and, along with the AGM-84, its air-breathing propulsion system.

One program attribute that is unquestionably important in determining the length of the development effort is the number of missiles launched during flight test. Data in Tables 2 and 4 enumerate the number of development missiles procured and the number launched during both the prototype and FSD flight test programs. In our description of test missiles, we use the term "prototype missile" to describe missiles procured as a part of pre-FSD development and the term "development missile" to describe those procured in support of the FSD program. Various nomenclature were used within and among the different pro-grams to describe test assets; we use these two standard terms. The number of test missiles procured is listed by contract, while the number of test missiles launched is listed by test phase.

Test phase definitions also differ among the services and programs. Navy FSD programs are characterized by three test phases. The first phase, contractor demonstration test (CDT) or contractor test and evaluation (CTE), is when the contractor must demon-strate the

Table 1. Characteristics of Air-to-Air Programs

Program Characteristics	AIM-7F Sparrow	AIM-7M Sparrow	AIM-9L Sidewinder	AIM-9M Sidewinder	AIM-54A Phoenix	AIM-54C Phoenix	AIM-120 AMRAAM
Military Service	Navy	Navy	Navy	Navy	Navy	Navy	Air Force
Prime Contractor	Raytheon	Raytheon	Raytheon	Raytheon	Hughes	Hughes	Hughes
Prototype	No	Yes	Yes	No	No	No	Yes
Modification	Yes	Yes	Yes	Yes	No	Yes	No
Prototype Missiles	N/A	5/5	14	N/A	N/A	N/A	16/16
Prototype Launches	N/A	2/3	11	N/A	N/A	N/A	3/5
Development Missiles	FSD: 34 FY68: 65 FY71: 29 FY72: 100	FSD: 44	FSD: 126	FSD: 139	FSD: 37 TP/VE: 26 SAT/VE: 31 FY71: 69	FSD: 15 Pilot Prod.: 30	FSD: 122
	Total: 228	Total: 44	Total: 126	Total: 139	Total: 163	Total: 45	Total: 122
Development Launches	CDT: 7 NTE: 42 OPEVAL I: 25 OPEVAL II: 25	CTE: 13 JTE: 22	CTE: 10 JET: 20 IOT&E/ OPEVAL: 30	CTE: 10	CDT: R&D, 24 TP/VE, 26 NTE: 11 OPEVAL: 24	CDT: 10 NTE: 6 OPEVAL: 15	DT&E/ IOT&E: 91
	Total: 99	–	Total: 60	–	Total: 85	Total: 31	Total: 91
Launch Platforms Used During FSD	F-4B F-4J F-4E	–	F-4J F-4E F-14A A-7	–	F-111B F-14A	F-14A	F-16C F-15C F-14A F/A-18

Note: N/A mean not applicable. Dashes (–) mean data were not available.

Table 2. Characteristics of Air-to-Air Missiles

Missile Characteristics	AIM-7F Sparrow	AIM-7M Sparrow	AIM-9L Sidewinder	AIM-9M Sidewinder	AIM-54A Phoenix	AIM-54C Phoenix	AIM-120A AMRAAM
Primary Targets	Fighters, bombers	Fighters, bombers	Fighters	Fighters	Bombers, cruise missiles	Bombers, cruise missiles	Fighters, bombers
Guidance Type	Semi-active radar	Semi-active radar	Passive infrared	Passive infrared	Active radar (semi-active midcourse)	Active radar (semi-active midcourse)	Active radar (command midcourse)
Length (ft.)	12	12	9.5	9.5	13	13	12
Diameter (in.)	8	8	5.0	5.0	15	15	7.1
Total Weight (lbs.)	510	510	190	190	985	1,000	342
Guidance Weight (lbs.)	63.9	61.6	26	26	166.5	179.0	89.0
Missile Cross-Section ($in.^2$)	50.2	50.2	19.6	19.6	176.6	176.6	39.6
Guidance Weight/Cross-Section ($lb./in.^2$)	1.27	1.23	1.32	1.32	.94	1.01	2.25
Total Impulse (lbs. x sec.)	36,073	31,000	13,912	13,912	97,120	97,120	_[a]

[a]Classified.

Table 3. Characteristics of Air-to-Surface Programs

Program Characteristics	AGM-65A TV Maverick	AGM-65D IIR Maverick	AGM-69A SRAM	AGM-84A Harpoon	AGM-86B ALCM	AGM-88A HARM	AGM-114 Hellfire
Military Service	Air Force	Air Force	Air Force	Navy	Air Force	Navy	Army
Prime Contractor	Hughes	Hughes	Boeing	McDonnell Douglas	Boeing	Texas Instruments	Rockwell
Prototype	No	Yes	No	Yes	Yes	Yes	Yes
Modification	No	Yes	No	No	No	No	No
Prototype Missiles	N/A	8	N/A	12	7	16	14
Prototype Launches	N/A	4	N/A	9	6	9	–
Development Missiles	FSD: 91	FSD: 33	–	WSD: 40 Pilot Prod.: 150	FSD: 24	Phase II: 27 Phase III: 45	FSD: 215
	Total: 91	Total: 33	–	Total: 190	Total: 24	Total: 72	Total: 215
Development Launches	Cat I: 15 Cat II: 37	DT&E: 14 IOT&E: 12	Cat I & II: 38	CTE/NTE: 33 OPEVAL: 33	DT&E: 10 Follow-on DT&E: 11	CTE: 18 NTE/IOT&E/ OPEVAL: 40	FSD: 169
	Total: 52	Total: 26	Total: 38	Total: 66	Total: 21	Total: 58	Total: 169
Launch Platforms Used During FSD	F-4D F-4EJ A-7	F-4D F-4E A-10A F-16A	B-52 FB-111	P-3C DE 1052 SSN-637 DLG/CAN DDG/ARM	B-52	A-7C A-7E F-4G A-6E	AH-64

Note: N/A mean not applicable. Dashes (–) mean data were not available.

Table 4. Characteristics of Air-to-Surface Missiles

Missile Characteristics	AGM-65A TV Maverick	AGM-65D IIR Maverick	AGM-69A SRAM	AGM-84A Harpoon	AGM-86B ALCM	AGM-88A HARM	AGM-114A Hellfire
Primary Targets	Armored vehicles	Armored vehicles	Deep air defense, fixed strategic	Surface ships	Fixed strategic	Ground radar	Armored vehicles
Guidance Type	Television	Imaging infrared	Inertial	Active Radar, Inertial Midcourse	Inertial with updates	Passive radar	Semi-active laser
Length (ft.)	8.2	8.2	14.0	12.5	20.8	13.6	5.4
Diameter (in.)	12.0	12.0	17.7	13.5	27.3	10.0	7.0
Total Weight (lbs.)	462.0	484.5	2,210	1,168	3,144	807	99.9
Guidance Weight (lbs.)	88.3	110.0	177	83	67	75	12.0
Missile Cross-Section (in.^{2})	113.0	113.0	245.9	143.1	585.1	78.5	38.5
Guidance Weight/Cross-Section (lb./in.^{2})	.78	.97	.72	58	.11	.96	.32
Total Impulse (lbs. x sec.)	5,426	5,426	251,392	N/A	N/A	_[a]	135.4

Note: N/A means not applicable.
[a]Classified.

basic capabilities of the missile. The second phase, Navy technical evaluation (NTE, sometimes referred to as TECHEVAL) or joint test and evaluation (JTE), is when the government determines the capabilities of the missile and decides whether it is ready for operational evaluation (OPEVAL). OPEVAL is when the government evaluates the missile in an operational environment. The three phases are generally serial in nature. For Air Force programs the phase analogous to CDT/CTE is Category I (Cat I) or development test and evaluation (DT&E). Air Force initial operational test and evaluation (IOT&E) or Category II (Cat II) phases have similarities with both the NTE/JTE and OPEVAL phases. Cat I and Cat II phases are associated with older Air Force programs and are serial in nature. DT&E and IOT&E generally overlap in combined test programs.

DEVELOPMENT PROGRAM SCHEDULES

This section summarizes schedule data both in tabular and graphical form. The data are presented in a manner consistent with the way they are analyzed. We present program schedule data characterizing development through production start and initial operational capability (IOC), including pre-FSD hardware prototyping.

Tables 5 and 6 present major milestones for the seven air-to-air- and the seven air-to-surface-missile programs, respectively. In order to compare across programs, milestone dates were normalized to the common milestone, FSD start. We defined FSD start as the beginning of FSD contract efforts, which usually corresponds to FSD contract award. FSD start was used because it represents the most unambiguous base point common to all programs; normalized milestones are expressed as months from FSD start. Other schedule interval data include time from prototype start to prototype first launch, FSD first launch to first production, and first production to initial operational capability. Also included are average intervals for each mission group.

The time from development start to the first guided launch is consumed by various activities. These include missile design, fabrication, and assembly of test hardware, and testing leading to initiation of the guided-launch program. This testing generally includes hardware integrated simulation, captive-missile flight test,

Table 5. Program Milestones and Intervals for Air-to-Air Programs

	AIM-7F Sparrow	AIM-7M Sparrow	AIM-9L Sidewinder	AIM-9M Sidewinder	AIM-54A Phoenix	AIM-54C Phoenix	AIM-120 AMRAAM	Average
Calendar Dates								
Prototype Start	N/A	8/76	6/71	N/A	N/A	N/A	2/79	N/A
First Launch, Prototype	N/A	–	4/72	N/A	N/A	N/A	6/80	N/A
FSD Start	7/66	4/78	7/72	2/76	12/62	9/77	12/81	N/A
First Guided Launch, FSD	3/68	4/80	10/73	2/78	5/66	5/80	5/85	N/A
First Production Delivery	1/76	1/83	3/78	–	3/73	8/82	9/88	N/A
IOC	4/76	1/83	5/78	9/82	12/73	12/86	10/89	N/A
Months From FSD Start								
Prototype Start	N/A	-20	-13	N/A	N/A	N/A	-34	-22.3
First Launch, Prototype	N/A	–	-3	N/A	N/A	N/A	-18	-10.5
FSD Start	0	0	0	0	0	0	0	0
First Guided Launch, FSD	20	24	15	24	41	31	41	27.4
First Production Delivery	113	57	68	–	123	59	81	83.7
IOC	117	57	70	79	132	111	94	94.3
Other Intervals (Months)								
Prototype Start to Prototype First Launch	N/A	20	10	N/A	N/A	N/A	16	15.3
FSD First Launch to First Production	94	33	53	–	82	28	40	55.0
First Prodction to IOC	3	0	2	–	9	52	13	13.2

Note: N/A means not applicable. Dashes (–) mean data were not available.

Table 6. Program Milestones and Intervals for Air-to-Surface Programs

	AGM-65A TV Maverick	AGM-65D IIR Maverick	AGM-67A SRAM	AGM-84A Harpoon	AGM-86B ALCM	AGM-88A HARM	AGM-114 Hellfire	Average
Calendar Dates								
Prototype Start	N/A	4/74	N/A	6/71	2/74	11/74	–	N/A
First Launch, Prototype	N/A	10/75	N/A	12/72	9/76	2/76	–	N/A
FSD Start	7/68	10/78	11/66	9/72	2/78	9/77	10/76	N/A
First Guided Launch, FSD	12/69	12/80	7/69	3/74	8/79	4/79	10/78	N/A
First Production Delivery	8/72	10/83	3/72	2/77	11/81	11/82	7/83	N/A
IOC	2/73	2/86	8/72	7/77	12/82	11/83	7/86	N/A
Months From FSD Start								
Prototype Start	N/A	-54	N/A	-15	-48	-34	–	-37.8
First Launch, Prototype	N/A	-36	N/A	3	-17	-19	–	-17.3
FSD Start	0	0	0	0	0	0	0	0
First Guided Launch, FSD	17	26	32	18	18	19	24	22
First Production Delivery	49	60	64	53	45	62	81	59
IOC	55	88	69	58	58	74	117	74
Other Intervals (Months)								
Prototype Start to Prototype First Launch	N/A	18	N/A	18	31	15	N/A	20.5
FSD First Launch to First Production	32	34	32	35	27	43	57	37
First Prodction to IOC	6	28	5	5	13	12	36	15

Note: N/A means not applicable. Dashes (–) mean data were not available.

and non-guided launches, including separation/jettison testing. The length of the interval from first guided launch to the first production delivery is driven mainly by the length of the test program. This, in turn, is determined by the number of missiles launched and the rate at which test launches are completed. Test program length is treated in more detail in the next subsection. Another factor is the degree of concurrency between the test program and the start of production; data on early production for selected programs is presented in subsection E.

The relationship between production start and IOC is less clear. For most programs, this interval was between 0 and 15 months. Noticeable exceptions are the AGM-65D, AGM-114A, and AIM-54C, which all took over 25 months. Because definitions of IOC differ across programs, and because these differences are not easily characterized, analysis related to this milestone was limited. Figures 2 and 3 display FSD milestones as expressed in months from FSD start for air-to-air and air-to-surface programs. Averages for all programs within each classification are plotted on the horizontal axes and values for individual programs are plotted on the vertical axes. Points above the 45-degree line represent values higher (longer time intervals) than the group average for a given milestone; those below represent values lower (shorter time intervals) than the average. Dispersion around the mean is greater for mile-stones that occurred later in the programs.

Looking at variations in schedule intervals across programs in the context of program and missile characteristics, we can see that certain patterns emerge. On average, schedule intervals for air-to-air programs were longer than those for air-to-ground programs. This is particularly evident for the interval from first guided launch to first pro-duction delivery. This interval was also longer for programs in which a large number of test missiles were launched. Missiles with heavier and more complex guidance systems tend to take longer to develop, both to first flight and to first production. We statistically test these and other schedule drivers in the next subsection.

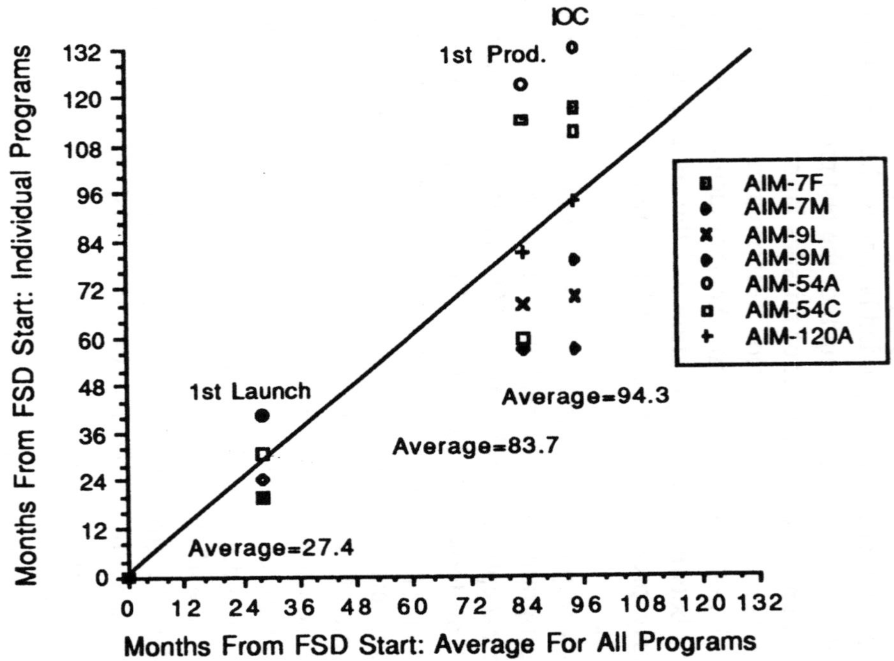

Figure 2. Full Scale Development Milestones for Air-to-Air Programs

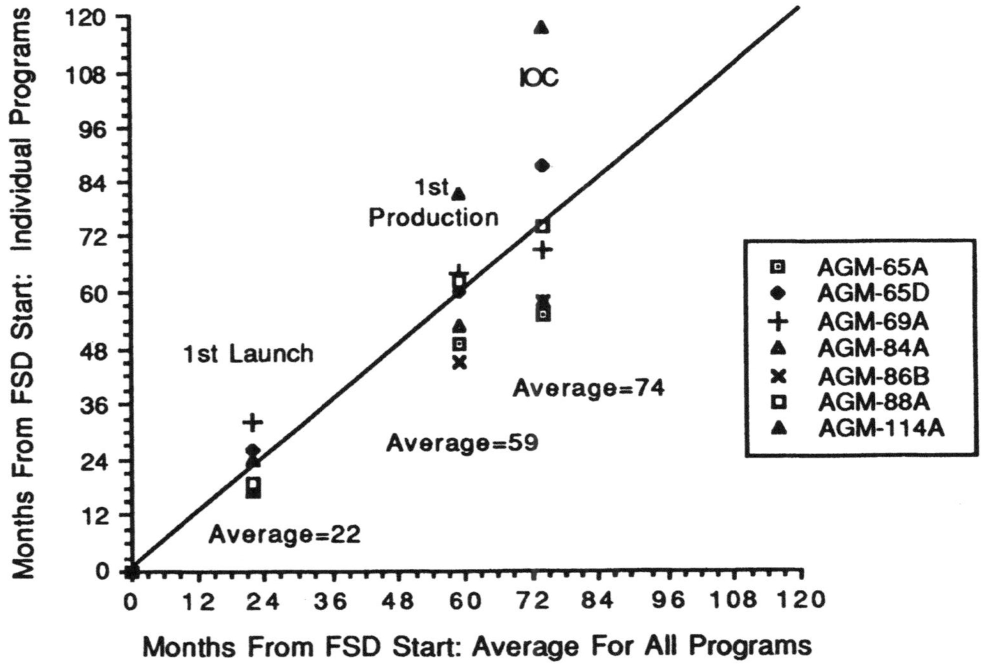

Figure 3. Full Scale Development Milestones for Air-to-Surface Programs

FLIGHT TEST PROGRAM

Table 7 summarizes test program data for thirteen of the programs in our sample. The only program for which no test data were available was the AGM-114. We present data by program and phase, including test start and end dates and the number of missiles fired. For most programs, we have a complete accounting of the test phases comprising the FSD launch program. We also have data on four of the seven pre-FSD prototype test programs. The variable of most interest is the launch rate. The launch rate in Table 7 was calculated as follows:

$$\text{Launches per month} = \left[\frac{\text{Number of test launches} - 1}{\text{Test duration in months}}\right],$$

where the test duration was the number of months separating the first and last launches, hence the subtraction of one launch from the numerator. The length of the individual test phases is determined by the number of missiles tested and the rate at which they were fired. One source of much schedule optimism at program start is the over-estimation of launch rates. We have also seen that major technical problems in development may require increases in the number of development missiles tested and subsequent program schedule stretches. In our analyses, emphasis was on explaining the variability in launch rates between programs because the latter problem does not lend itself to prediction.

Certain patterns in the variability of launch rates become apparent when examining the data in Table 7. Test phases that occurred later in the program had higher launch rates. Air-to-air programs had lower rates than air-to-surface programs. The two programs that had a substantial number of launches from surface ships, the AIM-7M and the AGM-84A, experienced higher rates than otherwise comparable programs. Programs that used multiple test sites also had higher rates. We test these and other possible determinants of launch rate in the data analysis section.

Table 7. Test Program Summary

Program	Phase	Number of Sites	Number of Launches	Test Start	Test End	Test length (Months)	Launches per Month
AIM-7F	CDT	1	7	3/15/68	11/6/68	7.8	.77
	NTE	1	42	12/10/69	2/2/72	25.8	1.59
	OPEVAL I	2	25	5/4/72	12/29/72	7.9	3.05
	OPEVAL II	2	25	8/73	9/74	13.0	1.84
AIM-7M	CTE	1	13	4/2/80	8/27/80	4.8	2.48
	JTE	1	22	8/27/80	2/24/81	6.0	3.53
AIM-9L	Pre-FSD	1	11	4/72	7/73	15.0	.67
	CTE	1	10	10/73	5/74	7.0	1.29
	JTE/OPEVAL/IOT&E	2	50	8/74	12/75	16.0	3.06
AIM-9M	CTE	1	11	4/72	7/73	14.9	.60
AIM-54A	CDT (FSD)	1	26	5/12/66	12/24/69	43.5	.58
	CDT (TP/VE)	1	27	3/9/70	10/19/73	43.4	.60
	NTE	1	11	6/5/73	6/27/74	12.7	1.18
	OPEVAL	1	24	8/23/74	7/13/76	22.7	1.01
AIM-54C	CTE	1	10	4/17/80	3/31/82	23.4	.38
	NTE	1	6	5/82	11/82	6.0	.83
	OPEVAL	1	15	3/83	6/84	15.1	.93
AIM-120A	DT&E/IOT&E	3	91	5/14/85	6/15/89	49.1	1.83
AGM-65A	Cat I	1	15	12/18/69	11/30/70	11.4	1.23
	Cat II	1	37	2/10/71	9/8/71	6.9	5.21
AGM-65D	DT&E/IOT&E	1	26	12/4/80	8/17/82	20.4	1.19
AGM-69A	Cat I/Cat II	1	38	7/29/69	7/7/71	23.3	1.59
AGM-84A	Pre-FSD	1	12	12/20/72	12/10/73	11.7	.94
	CTE/NTE	1	33	3/31/74	6/27/75	14.9	2.15
	OPEVAL	1	33	8/1/75	3/1/77	19.0	1.68
AGM-86B	Pre-FSD	1	3	9/9/76	11/30/76	2.7	.74
	DT&E	1	10	8/3/79	1/22/80	5.7	1.59
	DT&E (follow-on)	1	11	6/10/80	5/5/81	10.8	.92
AGM-88A	Pre-FSD	1	9	2/25/76	6/22/77	15.9	.50
	CTE	1	18	2/28/79	10/31/80	20.1	.85
	NTE/OPEVAL/IOT&E	2	40	8/6/81	10/5/82	14.0	2.79

EARLY PRODUCTION

Data were collected characterizing the early production periods for a portion of our sample programs. Table 8 presents production milestones and other data describing early procurement lots for ten programs. Included are data for both pilot production and production lots. Data for one-second source program, the AIM-7F, are also included.

Table 8. Early Production Experience

System	Description	Quantity	LL	FF	First Delivery	Time in Months: LL to First Delivery	FF to First Delivery	Test End To LL	Test End To FF
AIM-7F	FY68 Pilot Production	65	3/68	6/69	7/70	28	13		
	FY71 Pilot Production	29	–	9/70	12/71	–	15		
	FY72 Pilot Production	100	–	6/72	8/73	–	14		
	FY73 Pilot Production	150	–	5/73	8/74	–	15		
	FY75 Production	600	2/74[a]	11/74	1/76	23[a]	14	-7[a]	2
AIM-7F (second source)	FY74 Qualification Lot	15	7/74	–	3/76	20	–		
	FY75 Pilot Production	70	8/75	–	6/77	22	–		
AIM-9L	FY76 Production	1,534	4/76	12/76[a]	3/78	23	14	4	13
AIM-54A	FY72 Production	240	12/70	12/71	3/73	27	15	-67	-55
	FY73 Production	180	–	12/72	2/74	–	14		
	FY74 Production	284	–	12/73	2/75	–	14		
AIM-54C	Pilot Production	35	9/79	–	10/81	25	–		
	FY80 Production	60	9/79	11/79	8/82	35	33	-57	-55
	FY81 Production	60	6/81	10/81	7/83	25	21		
	FY82 Production	72	6/82	10/82	6/84	24	20		
AIM-120A	FY86 Production	180	11/86	10/87	9/88	22	11	-31	-20
	FY88 Production	400	12/87	7/88	8/89	20	13		
	FY89 Production	1,270	10/88	7/89	8/90	22	13		
AGM-65A	Option A Production	2,000	2/71[a]	7/71	8/72	18[a]	13	-7[a]	-2
	Option B Production	5,000	–	11/72	1/74	–	14		
	Option C Production	10,000	–	10/73	11/74	–	13		
AGM-65D	FY82 Production	200	4/82	9/82	10/83	18	13	-4	1
	FY83 Production	900	–	4/83	11/85		31		
	FY84 Production	1,980	4/84	4/85	8/86	28	16		
AGM-69A	FY71 Production	122	6/70	1/71	3/78	21	14	-13	-6
AGM-84A	Pilot Production	150	1/74	7/74	8/75	19	13		
	First Production	345	6/75	2/76	2/77	20	12	-21	-13
AGM-86B	FY80 Production	225	10/79	3/80	11/81	25	20	-19	-14
	FY81 Production	480	10/80	–	11/82	25	–		
	FY82 Production	440	10/81	–	10/83	24	–		
AGM-88A	First Production	80	1/81	12/81	12/82	23	12	-21	-10

Note: Dashes (–) mean data were not available.
[a]Estimates.

Production times were measured as both the number of months from long-lead release (LL) and full-funding release (FF) to the first missile delivery for a given lot of missiles. The full-funding release date generally corresponds to the contract award date. For the first production lot in each program, we included measures of concurrency. The measures were the spans of time between the end of the test program (defined as the end of the OPEVAL, IOT&E, or Cat II phase) and the long-lead release and full-funding release milestones. The more overlap between the test programs and production (concurrency), the smaller the values of the two measures. In order to derive a full set of concurrency measures, we estimated dates for long-lead release for the AIM-7F and AGM-65A programs and full-funding release for the AIM-9L program. In estimating a date for the AIM-9L program, we added the average time from long-lead release to full-funding release for the sample (9 months as calculated in the data analysis section) to the known date for long-lead release. For the AIM-7F, we subtracted this average from the known full-funding release date. For the AGM-65A, we subtracted the analogous time span for the AGM-65D program.

Certain outliers and anomalies exist in the data. The longest production time intervals were for the first production lot of the AIM-54C program. Long-lead release for this lot occurred on the same date as for the pilot production lot. Another unusually long production time was for the AGM-65D's FY83 production lot. Outliers for the concurrency measures include both AIM-54 development programs. Because of the stretch in the AIM-54 program due to the F-111B cancellation, a large number of test launches were accomplished before OPEVAL, many of which resembled operational tests; consequently, production decisions were made long before OPEVAL even started. In the case of the AIM-54C, the first three production lots were relatively small, making them similar to pilot production lots, and these lots were therefore contracted for early in the program.

Looking across the values for the majority of the programs, we see that production times from long-lead and full-funding release fall between 18 and 28, and 11 and 21 months, respectively. There seems to be no pattern in the variability of the production times. Examining the measures of concurrency (disregarding the AIM-54 programs), we find four programs that are essentially non-concurrent and four that are concurrent. For the four non-concurrent programs, the AIM-7F, the AIM-9L, the AIM-65A, and the AIM-65D, full-funding and long-lead release occurred after or slightly before test completion. For the four concurrent programs, the AIM-120A, the AGM-84A, the AGM-86B, and the AGM-88A, long-lead and full-funding release occurred between 19 and 31, and 10 and 20 months, respectively, before test end. Of these last four programs, the AIM-120A shows the most concurrency. The AGM-69A is on the borderline between the two classes of programs. Our analyses of production data, discussed in the next section, explored other ways of characterizing concurrency.

DATA ANALYSIS OVERVIEW

The four program intervals we analyzed are: (1) time to first guided-launch, (2) length of development flight test, (3) early production times, and (4) program length from first guided-launch. The relationships between the four intervals are depicted in Figure 1 using the AIM-120A program as an example. First guided launch marks the culmination of early design and manufacturing activity and initiates the guided-launch flight test program. During flight testing, information is gathered to be used both in refining the missile design and in supporting production decisions. The concurrency of development and production can be seen in the overlap of development flight testing and early production activity. This production activity is characterized by production times for the early procurement lots, expressed as the interval between long-lead and full-funding release and the delivery of the first production missile. Originally, we wanted to develop a single equation to estimate the interval spanning total FSD program length as defined by the period from FSD start to the delivery of the first production missile. The equation would serve as an aggregate check on a total FSD length estimate derived from combining estimates for the other intervals. The drivers of schedule intervals before and after first guided launch are different enough that estimating a single equation characterizing the program as a unified whole is not practical or theoretically justifiable. Instead, we estimated the interval between first guided-launch and the first production delivery.

TIME TO FIRST GUIDED LAUNCH

In developing a regression model to estimate the number of months required from FSD start to first guided launch (TFGL), we tested many different specifications. Attempts to use missile characteristics as explanatory variables were fruitful. Of particular merit were variables characterizing guidance size and complexity. The guidance system is generally a missile's highest-value component and the subject of the most technological advance. As such, development of the guidance system typically sets the pace of a missile's development. Guidance system parameters used include guidance weight (GWt) and a measure of guidance system packaging complexity. Obviously, weight is not an ideal measure of guidance complexity because a major goal in newer, more advanced systems is the miniaturization of components. If we were focusing on a narrow class of missiles, weight probably would not explain much variation in time to first guided launch. However, our data base is diverse in that guidance weights range from 26 to 179 pounds. In addition to guidance weight, we developed a measure to serve as a proxy for guidance system packaging complexity (GCPLX). This measure was expressed as:

$$\text{GCPLX} = \left[\frac{\text{Guidance weight (lbs.)}}{\text{Missile cross-section (sq.in.)}}\right] .$$

We would have preferred a measure of guidance density, but complete and consistent data on guidance volume were not available.

Another missile characteristic variable that seemed to be related to schedule was the maximum velocity of the missile as measured by its Mach number (MACH). Other variables of interest included 1/0 indicator variables designating prototype (PROTO) and modification programs (MOD), and whether the missile had an air-to-air (AA) or air-to-surface mission. We also formulated a time-trend variable, defined as the number of months from January 1940 to first guided launch (TIME)

One problem with the missile characteristic parameters is that certain parameters are very closely related to one another. In statistical terms, they are collinear. Table 9 presents a correlation matrix of independent variables.

Table 9. Correlation Matrix of Independent Variables

	GWt	GCPLX	MACH	AA	PROTO	MOD	TIME
GWt	1.00						
GCPLX	-.35	1.00					
MACH	.55*	.40	1.00				
AA	-.18	.64**	.64**	1.00			
PROTO	-.42	.25	-.33	-.17	1.00		
MOD	-.35	.18	-.03	.51*	.00	1.00	
TIME	-.23	.58**	.08	.16	.54*	.29	1.00

**Significant at the .05 level.
*Significant at the .10 level.

From the table, we can see that GCPLX is highly positively correlated with MACH, AA, and TIME, where MACH and AA are correlated with each other. These relationships are consistent with intuition. Air-to-air missiles are required to reach high velocities to intercept high-speed targets and are likely to have their cross-sectional area minimized for a given guidance package weight. As for the relationship between GCPLX and TIME, the overall trend over time has been toward the miniaturization of components. For example, if we look across time at medium-range air-to-air missiles, we see the AIM-7F had a guidance system with a weight of 64 pounds packaged in a missile with an 8-inch diameter, while the modern AIM-120A has a guidance system with a weight of 89 pounds packaged in a missile with a 7.1-inch diameter.

The correlation between MOD and AA is a result of our sample composition. Five of the six modification programs involved air-to-air missiles. The one air-to-surface modification program, the AGM-65D, is more closely related to a new start than the other modification programs. The AGM-65D's imaging infrared guidance system is of a totally different type than the TV system of the AGM-65A. Modifications of the air-to-air systems in our sample involved extending the capabilities of and repackaging existing guidance system technologies. The sample can be characterized as being made up of three types of developments, new air-to-surface systems, new air-to-air systems, and modified air-to-air systems. The significant positive correlation between TIME and PROTO may be indicative of the trend toward prototyping as an acquisition strategy.

Given the interrelationships in the data, we found two core independent variables, guidance weight and guidance complexity, to be the best predictors of time to first guided launch. The resulting equation and measures of statistical significance and model fit are:

$$TFGL = -4.0 + \underset{(.01)}{.15\,(GWt)} + \underset{(.01)}{14.1\,(GCPLX)}$$

$N = 12$ $\quad R^2 = .82$ $\quad$ Adjusted $R^2 = .78$ $\quad SEE = 4.1$

Significance levels are in parentheses below the parameter estimates. N is the number of observations and SEE is the standard error of the estimate. The regression results show all parameter estimates significant at the .01 level. Only twelve of our fourteen programs were included in the data sample used to estimate the equation. The AGM-86B and AGM-114A were not included in the sample. The AGM-114A was a large outlier with values for GWt and GCPLX much lower than for the rest of the sample, although its time to first guided launch of 24 months was close to the average. The target acquisition system associated with the AH-64/AGM-114A was developed in parallel with the missile and the delayed availability of AH-64 launch platforms negatively affected the program's schedule. Although the AGM-86B was not an outlier when it was included in the regression, we think that the variables in the equation are not good proxies for the schedule drivers for that program. Figure 4 plots time to first guided launch predicted by the above equation against program actuals. Values above the 45-degree line are overestimated by the model; values below that line are underestimated.

Obvious outliers include the AIM-54A and AIM-54C programs. The AIM-54A and AIM-54C share similar guidance system parameters with slightly higher values associated with the AIM-54C. Consequently, the values predicted for the two AIM-54 programs are similar despite the fact that the AIM-54C took ten months less time to first guided launch than the AIM-54A. The primary difference between the two programs was that the AIM-54C was a modification program. When the MOD variable was included in the

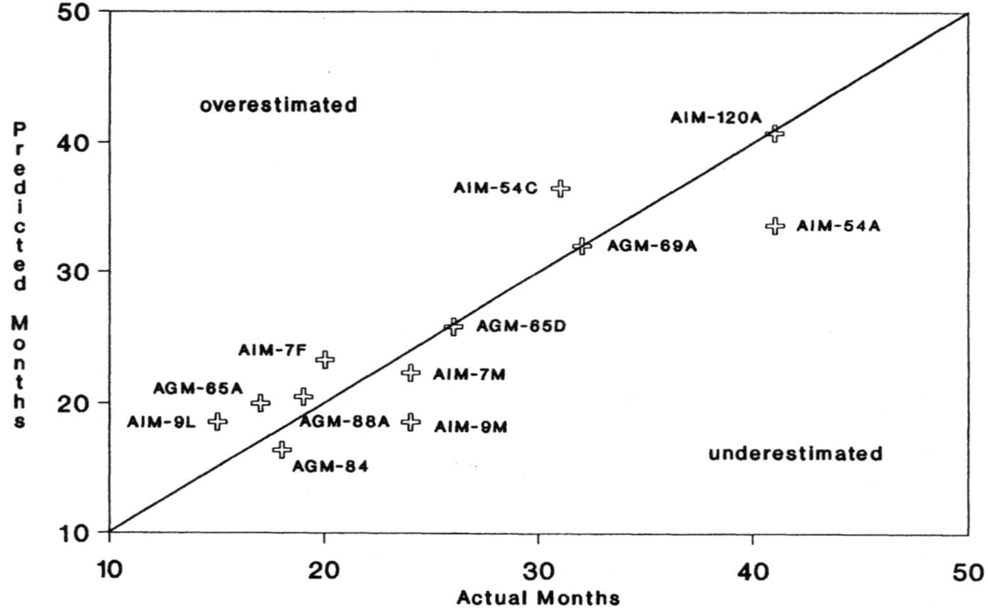

Figure 4. Predicted Vs. Actual Time From FSD Start to First Guided Launch: Two-Variable Model

equation, it was not statistically significant. We formulated another indicator variable that distinguishes air-to-air programs that were new starts (NEW_AA) from the rest of the sample. The equation with this variable included is:

$$TFGL = 4.6 + .11\ (GWt) + 8.1\ (GCPLX) + 9.5\ (NEW\_AA)$$
$$(.01) \qquad (.04) \qquad (.03)$$

N = 12 $\quad R^2 = .91 \quad$ Adjusted $R^2 = .87 \quad$ SEE = 3.2

The regression results show all parameter estimates significant at the .05 level. Figure 5 plots time to first guided launch predicted by the above equation against program actuals. The fit is improved significantly from the two-variable model; the AIM-54A and the AIM-54C are no longer outliers.

A weakness of the NEW_AA variable is that only two new air-to-air programs are in the sample, the AIM-54A and AIM-120A. We tried a second indicator variable to differentiate the modified air-to-air programs from the air-to-surface programs, but its parameter estimate was not significantly different from zero. The interpretation drawn from the analyses is that for the three types of systems in the sample, time to first guided launch is similar for new air-to-surface programs and modified air-to-air systems (given similar guidance attributes), but is longer for new air-to-air systems.

FLIGHT TEST

Our approach to estimating flight test duration consisted of two steps. First, regression analysis was employed using data presented in Table 8 to estimate launch rates. Each launch-rate value was associated with a development program and a test phase. Given estimated values for launch rate, flight test program length can be derived from the following relationship:

$$\text{Test phase duration in months} = \left[\frac{\text{Number of test launches} - 1}{\text{Launch rate per month}}\right],$$

where the required number of test launches is taken as a variable decided upon outside the model. We also estimated an equation for

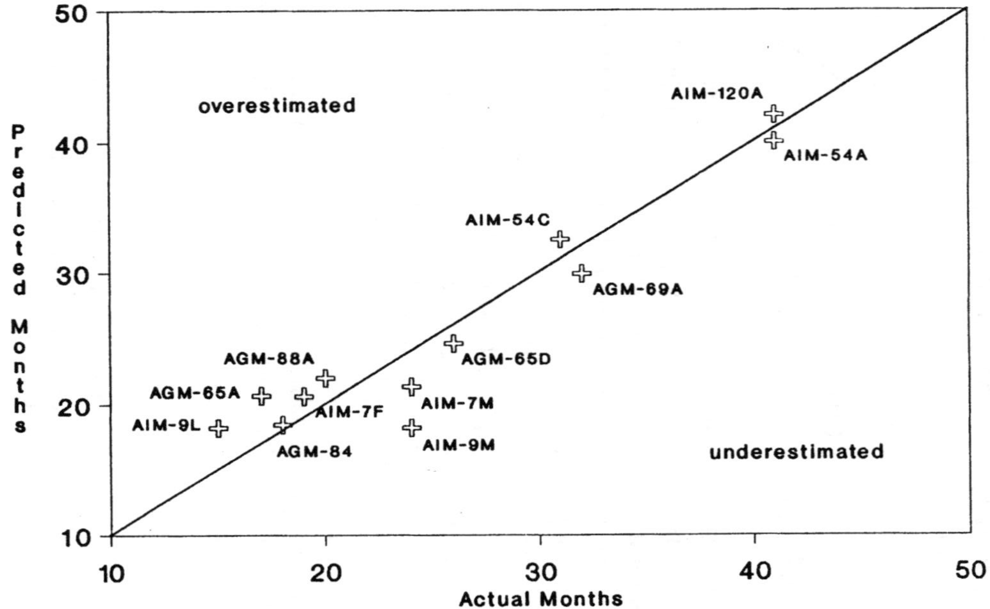

Figure 5. Predicted Vs. Actual Time From FSD Start to First Guided Launch: Three-Variable Model

the number of test missiles, to be used as a check against the proposed number of test launches.

We found that much of the variation in launch rates can be explained by technological and program variables. These include guidance section weight, whether the missile had an air-to-air or air-to-ground mission, what test phase the launches occurred in, the number of test sites, and whether a substantial number of launches were performed from surface ships.

Guidance weight was used as a proxy for the technical complexity of the hardware being tested. The hypothesis was that the more complex the hardware being tested, the more time it would take to prepare for each launch event. Launch rates were expected to be lower for air-to-air missiles than for air-to-surface missiles. Air-to-air testing is generally more difficult because the availability and preparation of air targets are fundamentally more problematic than surface targets, and air-to-air scenarios are generally more complex. In addition, air-to-air missiles tend to have a higher degree of hardware complexity.

Launch rates were also expected to be higher for test phases later in the program. Earlier test phases were characterized by more intense data reduction and analysis activities. Also, the relative maturity of development hardware was higher for test phases later in the program. For most programs, there were multiple data points each representing a test phase. Air Force and Navy test programs do not follow the same structure, so some interpretations were made when classifying test phases into global categories. After examining the data and applying our knowledge about the nature of Navy and Air Force test programs, we classified the test phases into three categories. In the data, these categories are represented by 1/0 indicator variables. The three categories are: pre-FSD (or prototype) testing (PFSD); development test and evaluation (DT&E), including Navy CDT and CTE and Air Force Cat I and DT&E; and initial operational test and evaluation (IOT&E), including Navy NTE, JTE, and OPEVAL and Air Force Cat II and IOT&E. In some programs, development testing and initial operational testing were performed in combined programs. We created a fourth variable, JOINT to characterize these programs. An argument could be made that Navy NTE/JTE test phases should be classified as DT&E phases instead of IOT&E phases; however, the data show that in terms of launch rate, NTE/JTE are more closely aligned to OPEVAL than to CDT/CTE.

We also considered two variables related to the test environment. The first was the number of major test sites at which launch testing took place (NSITES). In some programs, a small number of specialized tests took place at separate sites. In our data base, these are not considered additional test sites. In one program, the AGM-65D, two test sites were used but were not employed at the same time, so we classified the program as having only one test site. The second variable identifies the programs (AIM-7M and AGM-84A) which had test launches performed from surface ships (SHIP). The data show higher rates for these programs, particularly in the case of the AIM-7M.

Regression results are consistent with our observations. We used a multiplicative specification, where launch rate per month (LRATE) is a function of guidance weight (GWt), mission (the indicator variable AA, which takes on the value of 1 for air-to-surface systems and 0 for air-to-air systems), and test phase (where there are a series of 1/0 indicator variables identifying test phases where IOT&E is the base case). We estimated the coefficients for this equation by first transforming the equations to a log-log form and then applying standard linear regression techniques. When this is done, the constant term of the equation becomes biased when it is transformed from the log-log form back to a multiplicative form. We adjusted for this bias by adding one-half of the regression mean square error to the constant term of the log-log equation. This adjustment is an approximate correction for the bias introduced by the log-log estimation methodology. The resulting equation is:

$$LRATE = 12.7\,\underset{(.01)}{(GWt)^{-.36}}\,\underset{(.04)}{(NSITES)^{.54}}\,\underset{(.01)}{e^{(SHIP\,.80)}}\,\underset{(.01)}{e^{(AA\,-.52)}}$$

$$\underset{(.01)}{e^{(PFSD\,-1.35)}}\,\underset{(.01)}{e^{(DT\&E\,-.78)}}\,\underset{(.03)}{e^{(JOINT\,-.50)}}$$

$N = 30 \quad R^2 = .81 \quad \text{Adjusted } R^2 = .75 \quad SEE = .32$

The signs of the coefficients are intuitively correct. The ordering of rate effects for the test phases is as expected with rates increasing for phases later in the program. The combined DT&E/IOT&E variable (JOINT) shows a rate effect between those for DT&E and the IOT&E base case. All coefficients are significant at the .05 level. The SEE may be converted into minus/plus percentages by the relationships (e^{-SEE}) - 1 and (e^{+SEE}) – 1.

Figure 6 plots launch rates predicted by the above equation against program actuals. Table 10 summarizes the prediction errors associated with fitting the equation to the data. Notable outliers include the AIM-9M CTE, the AGM-65 Cat II, and the AGM-84A CTE/NTE.

In addition to estimates of launch rates, estimates of the number of missiles to be launched and their distribution across test phases are needed in order to estimate test program lengths. The number of launches required to test a missile's readiness for production and deployment is based on engineering analyses made prior to the start of FSD. Sometimes the estimates are raised or lowered in the course of development. For example, the number of test launches for the AIM-7F increased from the original plan because of technical problems, and the number of launches for the AIM-65D were decreased because of funding constraints. Examination of the number of test launches by FSD program reveals certain patterns. The programs with the greatest number of launches (excluding the AGM-114A) were air-to-air developments. The two air-to-air missile developments that were new starts, the AIM-54A and AIM-120A, had 85 and 90 launches, respectively. Total launches for new tactical air-to-surface missile developments (again excluding the AGM-114A) fell within a narrow range: 52 launches for the AGM-65A, 58 launches for the AGM-88A, and 66 launches for the AGM-84A. The AGM-114A, which had 169 launches, is a conspicuous outlier. This large number of launches may be a result of the low unit cost of the AGM-114A when compared with the other missiles. For most programs, the high unit cost of test assets requires that the number of launches be minimized given the amount of information needed to be gathered. The two strategic systems, the AGM-69A and AGM-86B had fewer test launches than the tactical systems with 38 and 21 launches each. The strategic systems made use of fewer types of launch platforms during testing than the other air-to-surface systems.

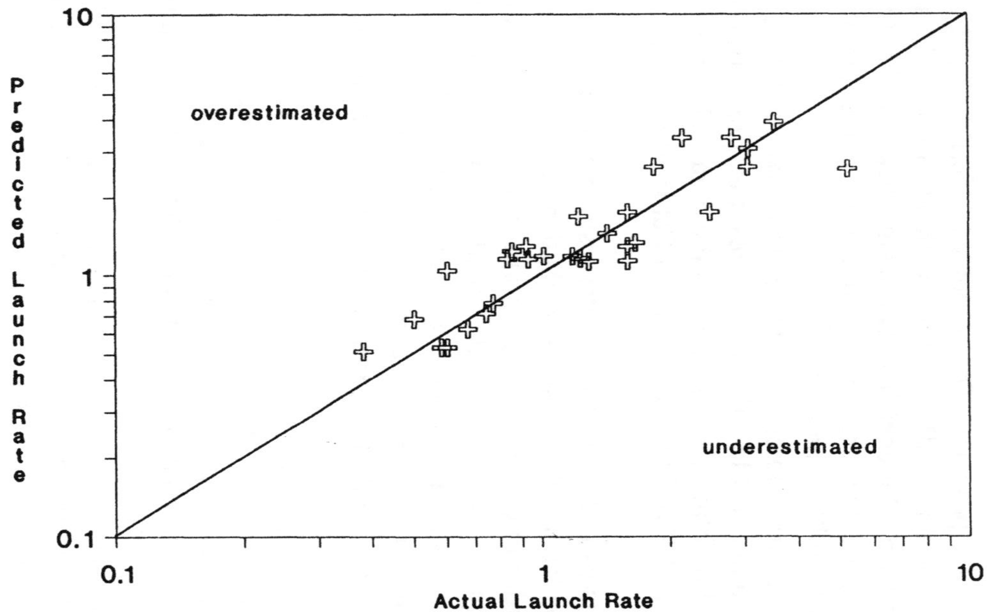

Figure 6. Predicted Vs. Actual Launch Rate per Month

Table 10. Launch Rate Prediction Error Summary

Program	Phase	Actual Launches/Month	Fitted Value	Residual	Relative Error
AIM-7F	CDT	.77	.77	.00	.00
	NTE	1.59	1.68	-.09	-.06
	OPEVAL I	3.05	2.43	.62	.20
	OPEVAL II	1.84	2.43	-.59	-.32
AIM-7M	CTE	2.48	1.72	.76	.30
	JTE	3.53	3.77	-.24	-.07
AIM-9L	Pre-FSD	.67	.60	.07	.10
	CTE	1.29	1.06	.23	.18
	JTE/OPEVAL/IOT&E	3.06	3.37	-.31	-.10
AIM-9M	CTE	.60	1.06	-.46	-.77
AIM-54A	CDT (FSD)	.58	.54	.04	.07
	CDT (TP/VE)	.60	.54	.06	.10
	NTE	1.18	1.19	-.01	-.01
	OPEVAL	1.01	1.19	-.18	-.17
AIM-54C	CTE	.38	.53	-.15	-.39
	NTE	.83	1.15	-.32	-.39
	OPEVAL	.93	1.15	-.22	-.24
AIM-120A	DT&E/IOT&E	1.83	1.63	.20	.11
AGM-65A	Cat I	1.19	1.14	.05	.04
	Cat II	5.21	2.50	2.70	.52
AGM-65D	DT&E/IOT&E	1.22	1.41	-.19	-.15
AGM-69A	Cat I/Cat II	1.59	1.18	.41	.26
AGM-84A	Pre-FSD	1.43	1.47	.04	-.03
	CTE/NTE	2.15	3.45	-1.30	-.61
AGM-86B	Pre-FSD	.74	.72	.02	.03
	DT&E	1.59	1.26	.33	.21
	DT&E (follow-on)	.92	1.26	-.34	-.37
AGM-88A	Pre-FSD	.50	.69	-.19	-.38
	CTE	.85	1.21	-.36	-.43
	NTE/OPEVAL/IOT&E	2.79	3.84	-1.05	-.38

There are also regularities for modification programs. For most modification programs, the number of test launches was considerably lower than was the case with comparable new programs. Exceptions were the AIM-7F, which had 99 launches, and the AIM-9L, which had 60 launches. Both missiles incorporated changes beyond modified guidance systems with major alterations in airframe (AIM-7F and AIM-9L) and propulsion (AIM-

9F) systems. The AIM-7F was also unusual in that unsatisfactory results during OPEVAL lead to the re-running of that 25-missile test phase. Modified missiles that did not have major changes to the missile's airframe and propulsion were the subject of fewer test firings, as evidenced by the AIM-54C (31 launches) and the AGM-65D (26 launches).

For both air-to-air and air-to-surface missiles there seems to be a relationship between the number of types of launch platforms used during testing and the number of test launches. As the number of platforms increases, the number of test launches should also increase – an important aspect of testing is the integration of the missile with the platform's fire-control system and airframe.

Given these observations, we estimated an equation to be used to check engineering estimates of the total number of test missiles launched during FSD (NMISS). In order to account for the regularities discussed previously, we regressed the total number of test launches against 1/0 indicator variables, identifying air-to-air missiles (AA) and modification programs (MOD) and a variable for the number of kinds of launch platforms used in FSD (NPLATFORM). The resulting equation was:

$$\text{NMIS} = 24.6\,(\text{NPLATFORM})^{.62}\, e^{(\text{MOD}\,-.43)}\, e^{(\text{AA}\,.79)}$$
$$(.01) \qquad (.04) \qquad (.01)$$

$N = 11 \qquad R^2 = .86 \quad \text{Adjusted } R^2 = .80 \qquad \text{SEE} = .24$

The signs of the coefficients are intuitively correct. Data points not in the equation are the AIM-7M, the AIM-9M, and the AGM-114A. Complete data on the AIM-7M and the AIM-9M were not available, and the AGM-114A was an outlier because of reasons already mentioned.

In plotting the regression errors against our time trend variable (TIME, the number of months from January 1940 to first guided launch) we found a negative correlation between the errors and TIME. This indicates that the equation underestimates the number of test launches for earlier programs and overestimates them for later programs. This is consistent with views expressed by contractor representatives who noted a decrease in the number of test launches required for more modern programs because of improved computer simulation. In accounting for this effect statistically, we estimated the above equation with the time trend variable. Although the parameter estimate for TIME is not significantly different from zero at traditionally accepted levels of significance, we chose to report the resulting equation because of the intuitive appeal of the variable. The equation is:

$$NMIS = 1365\ (NPLATFORM)^{.61}\ (TIME)^{-.67}\ e^{(MOD\ -.38)}\ e^{(AA\ .72)}$$
$$(.01) \quad (.15) \quad (.05) \quad (.01)$$

$N = 11$ $\quad R^2 = .90$ $\quad$ Adjusted $R^2 = .84$ $\quad$ SEE = .21

Figure 7 plots the number of launches predicted by the above equation against program actuals.

When we look at the distribution of test launches across phases, some patterns become evident. Pre-FSD phases were made up of the fewest tests, ranging between 3 and 11 launches. The number of launches in CTE, CDT, DT&E, and Cat I phases ranged from 10 to 20. The remainder of FSD launches took place during the NTE, JTE, OPEVAL, IOT&E, and Cat II phases. (In the AIM-7F program, 11 of 18 CDT launches were performed as joint CDT/NTE launches.) The exception to this pattern was the AIM-54A, where most test firings (50 of 85) took place during CDT. This was due to the special circumstances of that program. The original launch platform for which the missile was developed, the F-111B, was cancelled, providing more time for development testing before the replacement platform, the F-14A, was available.

EARLY PRODUCTION

In our analysis of early production data, the emphasis was on both the production times for the early procurement lots and the concurrency between production and development. Production times are as defined in Section II: the time from production-lot long-lead and full-funding release to the first lot delivery for the early production lots. We characterize concurrency by looking at the progress of the flight-test program in relation to production long-lead and full-funding milestones.

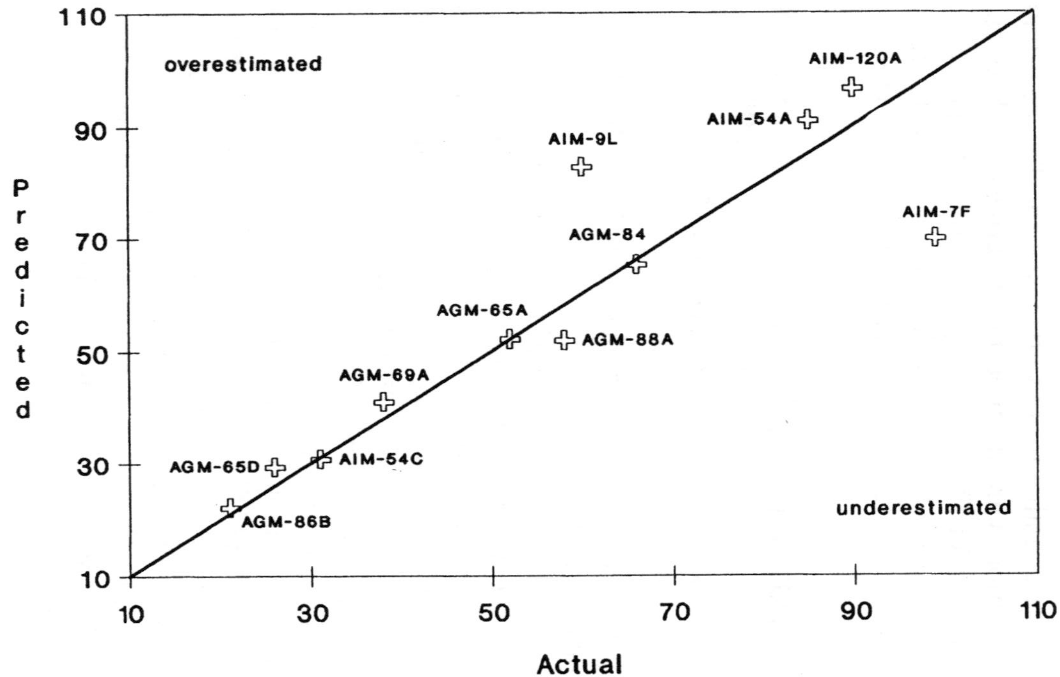

Figure 7. Predicted Vs. Actual Number of Missile Test Launches: Three-Variable Model

No statistical relationships between missile or program characteristics and production times were discovered. Neither did we find any relationship between the cumulative quantity produced and production times. Fortunately, there is little variability in the two intervals with the exception of the outliers mentioned in Section II, the AIM-54C's FY81 lot and the AGM-65D's FY83 lot. Because of this lack of variability, we think that simple descriptive statistics adequately characterize production times. Descriptive statistics for both production intervals are presented in Table 11. In calculating the statistics, data for the two anomalous production lots were removed from the sample.

Table 11. Descriptive Statistics for Production Times

	Long-Lead Release to First Production, N=21 (months)	Full-Funding Release to First Production, N=23 (months)
Minimum	18	11
Maximum	28	21
Range	10	10
Average	22.8	14.1
Standard Deviation	3.0	2.4

In developing alternative measures of concurrency, we looked beyond the time interval measures presented in Table 8 and examined the actual progress of the flight test program in relation to both long-lead and full-funding release. Table 12 lists the number of missile launches accomplished at the two production milestones and the percentage of total program launches they represented. Estimated values are based on estimated long-lead and full-funding dates.

These measures lead to somewhat different conclusions about the relative concurrency of the programs when compared with the simple interval overlap measures presented in Table 10. The AIM-54A was a highly concurrent outlier according to the interval measures. When looking at the percentage measures above, it emerges as a program of average concurrency. The low launch-rate and very long test period for the AIM-54A program meant that even at over 50 months prior to test end, over half of total launches had been completed. The AIM-54C remains an outlier; the two production milestones occurred before the first launch.

Table 12. Measures of Concurrency

First Production Lot	Launches at LL	Launches at FF	Total Launches	Percent at LL	Percent at FF
AIM-7F FY75 Production	86[a]	99	99	87[a]	100
AIM-9L FY76 Production	60	60	60	100	100
AIM-54A FY72 Production	33	49	85	39	58
AIM-54C FY80 Production	0	0	21	0	0
AIM-120A FY86 Production	19	43	91	21	47
AGM-65A Option A Production	16[a]	41	52	31[a]	79
AGM-65D FY82 Production	16	26	26	62	100
AGM-69A First Production	10	28	38	26	78
AGM-84A First Production	29	39	66	44	59
AGM-86B FY80 Production	4	10	21	19	48
AGM-88A First Production	19	25	58	33	43

[a]Estimates.

Three of the four programs that were unambiguously nonconcurrent according to the interval measures also show as nonconcurrent programs here. The exception is the AGM-65A. Although the estimated long-lead release date preceded the end of AGM-65A Cat II testing by only seven months, 36 launches were completed in that period owing to the very high launch-rate achieved in Cat II testing. The two programs with divergent measures of concurrency at long-lead and full-funding release, the AGM-65A and AGM-69A, are characterized by widely differing launch rates at the beginning and end of their test programs.

According to both percentage measures, three of the programs (excluding the AIM-54C) are shown to be the most concurrent, the AIM-120A, the AGM-86B, and the AGM-88A. The missiles developed during these programs had unique capabilities not approximated in inventory missiles. Conversely, missiles developed during the least concurrent programs, the AIM-7F and AIM-9L, were incremental improvements to existing systems.

The measures of concurrency presented in this section are important in estimating the interval from first guided launch to first production.

PROGRAM LENGTH FROM FIRST LAUNCH

We would have preferred to estimate total program length (defined as the time from FSD start to the delivery of the first production missile) to be used as an aggregate check on estimates derived for the other schedule intervals. Unfortunately, the determinants of time to first launch and time from first launch to first production are just too different. According to our regression work, time to first launch is a linear function of technological variables. Time from first launch to first production is a function of the number of missiles launched in flight test, the rate at which they are launched, the overlap between production start and flight test, and production time. The number of test launches is clearly unrelated to time to first launch even in a coincidental fashion. Although technological variables similar to those included in the time to first launch equation are also determinants of launch rate, our regression work shows that they affect launch rate nonlinearly.

In order to estimate a correctly specified equation for the number of months from first launch to first production delivery (FL_FDEL), we were forced to employ nonlinear least-squares regression. We characterized both flight test length and its concurrency with production by using the number of flight test missiles launched at long-lead release (NMISLL) as an independent variable. This variable interacts multiplicatively with missile characteristic variables that are determinants of launch rate. As production time has no variability related to any of these factors, it is best characterized by an additive intercept. The combination of an additive intercept with the multiplicative terms describing flight test length requires that we estimate the equation using nonlinear least squares. The resulting equation and measures of model fit are:

$$\underset{}{FL\_FDEL} = \underset{(.01)}{27.7} + .0115\,(NMISLL)^{\underset{(.05)}{1.02}}\,(GWt)^{\underset{(.02)}{.77}}\,e^{(AA\ \underset{(.01)}{.94})}$$

$$N = 11 \qquad R^2 = .97$$

Significance levels are based on single-tail t-tests using the asymptotic standard errors.

The parameter estimates are generally consistent with expectations. The additive intercept is close to the average time period from long-lead release to first production delivery, 22.8 months. The parameter estimate of the exponential term on the NMISLL variable is close to 1. This is consistent with the fact that the relationship between the number of missiles launched and test length tends to be linear for any given test phase. The increase in launch rate in later test phases is not evident in the effects of the NMISLL variable because for most programs long-lead release occurs during or shortly after the first test phase of the program. It was expected that the interval should increase with guidance weight and be longer for air-to-air programs. Figure 8 plots time from first guided launch to first production predicted by the above equation against program actuals.

The equation estimating program length from first launch has at least two uses. First, it can provide a quick method of assessing the total length of proposed FSD programs given an estimate of time to first guided launch. The other, and perhaps a more interesting use is that of a consistency check on schedule assessments produced from an integrated set of the two interval estimates addressed in the earlier sections of this paper.

SUMMARY AND CONCLUSIONS

In our analysis of air-launched missile schedules we have found that program length is affected by a variety of factors or schedule drivers. By decomposing program schedules into multiple intervals, we have been able to better identify and understand these schedule drivers.

Development programs for missiles with complex guidance systems tend to take longer than those involving simpler guidance systems for two reasons: complex guidance systems take longer to first guided launch and they experience lower launch rates during flight testing. Air-to-air programs tend to be of longer duration than air-to-surface programs because of lower launch rates during flight testing and the need to test-launch more missiles. The schedule advantage of a modification program over a new start manifests itself both in a shorter time to first guided launch (for air-to-air programs at least) and fewer missiles required in flight testing.

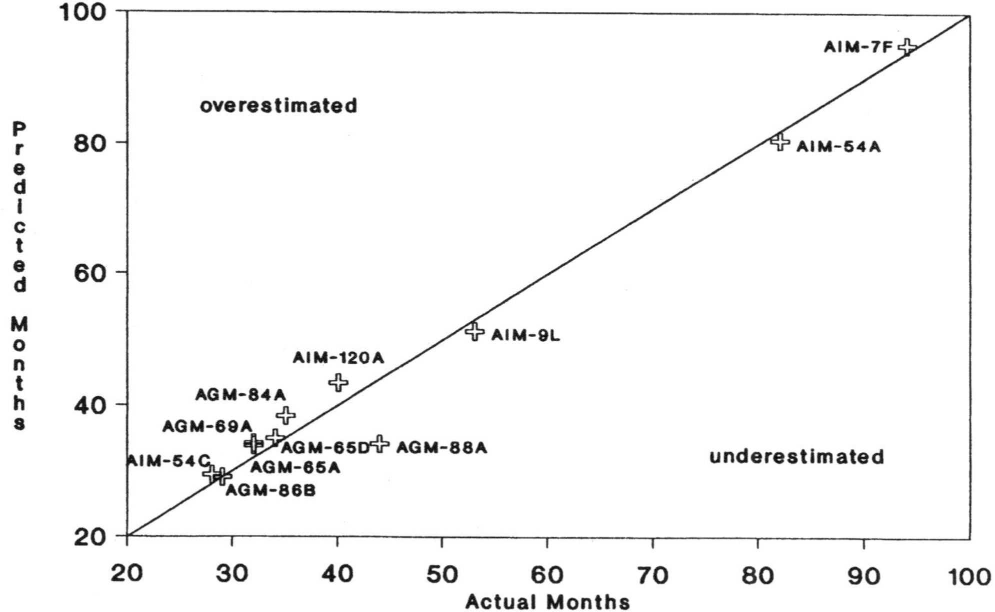

Figure 8. Predicted Vs. Actual Time From First Guided Launch to First Production

Although we were not able to explain any variation in early production times, we were able to characterize the degree of concurrency between development and production, which is a very important driver of overall development program length.

By quantifying these and other relationships, we think that we have created a useful set of tools for analysts to use in reviewing proposed schedules for air-launched missile programs. Because development program costs are often driven by test and schedule factors, these tools should also prove helpful to cost analysts.

References

[1] Harmon, Bruce R., Lisa M. Ward and Paul R. Palmer Jr., *Assessing Acquisition Schedules For Tactical Aircraft,* Institute For Defense Analyses Paper P-2105, February 1989.

[2] Rothman, M.B., *Aerospace Weapon System Acquisition Milestones: A Data Base*, RAND, October 1987

[3] Gogerty, David C., Bruce M. Miller and Paul R. Palmer Jr., *Acquisition of Contemporary Tactical Munitions, Volume II: Case Studies,* Institute For Defense Analyses Paper P-2173, June 1989.

[4] *Jane's Weapon Systems* 1987-1988, Jane's Publishing company Ltd., London, 1987

[5] *Aircraft Armament Volume II*, Interavia S.A., Geneva, Switzerland, 1982

[6] *DMS Market Intelligence Reports* (Missiles), DMS Inc., 1985

[7] Kluge, Arthur J., *The Tactical Missile Technical Data Handbook,* Tecolote Research, Inc. TM-102, 1977

[8] Jaffurs, C. Z., A. J. Kluge and D.C. Morrison, *Missile Cost Data Book,* Tecolote Research, Inc., November 1984.

[9] Eichblatt, Emil J., *Test and Evaluation of an Air-to-Air RF Guided Missile*, Pacific Missile Test Center, Point Mugu, California, May 1986.

[10] Friedman, Norman, *U.S. Naval Weapons*, Naval Institute Press, Annapolis, Maryland, 1983

[11] Gunston, Bill, *The Illustrated Encyclopedia of the World's Rockets and Missiles*, Salamander Books Ltd., London, 1979

[12] Nicholas, Ted G., *U.S. Missile Data Book,* Data Search Associates, October 1987

[13] York, James L., et al, *Tactical Missile Development Cost,* Science Applications International Corporation, May 1987.

[14] Vokolek, Alan S., and Damon C. Morrison, *Naval Weapons Center Cost Model*, Tecolote Research Company, CR-0171, January 1987.

[15] Luttrell, W.D, *Phoenix Missile System Research and Development Program*, Naval Missile Center, Technical Publication TP-70-51, June 1970.

[16] *Navy and Air Force Integrated Logistics Support Plan For Sidewinder AIM-9L Weapon System*, February 1977.

[17] *AIM-9L Independent Cost Analysis Summary Documentation,* Air Force Systems Command. August 1973.

[18] Dwyer, Thomas J. and Arthur J. Kluge, *Tactical Missile Hardware Cost Study, the Sparrow Missile.*,Tecolote Research, Inc. TM-57, July 1977

[19] *Sparrow III Master Plan Excerpts*, Raytheon Company, Lowell Massachusetts, 1973.

[20] Dwyer, Thomas J. and Arthur J. Kluge *Tactical Missile Hardware Cost Study, the Maverick Missile.*, Tecolote Research, Inc. TM-41, July 1977

[21] Geiger, Clarence J., *History of the AGM-69A SRAM: Short Range Attack Missile, 1963-1975, Volume I - Narrative,* History Office, Aeronautical Systems Division, Air Force Systems Command.

IV. Uncertainty in Cost Estimating

A General Analytic Approach To System Cost Uncertainty Analysis

Paul R. Garvey
The MITRE Corporation, Bedford, Massachusetts 01730

Abstract

This paper presents the foundations of a recently developed analytic approach to system cost uncertainty analysis. The approach is referred to as the Analytic Cost Probability (ACOP) model; and its structure is sufficiently general to meet the characteristics of any program definition. The analytic nature of the ACOP model facilitates the identification of cost variance drivers and measures their overall impact on the system cost. The ACOP model offers several techniques for treating correlation between cost elements of a work breakdown structure; a technical issue that has not been widely discussed in the literature or accounted for in existing models. An illustrative analysis using the ACOP model on a hypothetical system is presented, and the mathematical foundations of the model are provided so that the cost analysis community may review, comment on, and expand upon the approach within their organizations.

Executive Summary

The Analytic Cost Probability (ACOP) model was initially developed in 1987 to support a planning cost study of a large-scale information system. The theory and concepts behind the ACOP model are so general that they can be applied to a wide range of programs, and the approach has provided new insights into technical issues often ignored. For instance, the important technical issue of how to describe probabilistically the cost of a system if statistical correlation exists between the elements, has not been fully treated in the literature or accounted for in existing models. In this paper, methods to formally incorporate correlation between the elements of cost are fully developed; and conditions are identified which permit the system cost probability distribution to be specified analytically, even when all the elements are *not* independent. In addition, an analysis of historical program cost data is presented to provide insight into how the major cost elements of a system are actually correlated.

Although various cost uncertainty analysis techniques have been available to the industry for many years, among the contributions to the field provided by the ACOP model include:

- its general structure which enables the analyst to tailor the model to specific characteristics of the program definition, in addition to identifying n-categories of cost uncertainty and isolating the effect of any one category on the total cost risk;

- the methods provided for incorporating correlation between the cost elements of the system, and the development of a useful theorem (presented in appendices A and B) for bounding the effect of correlation on the system cost variance.

Thus, we now have a fully developed analytic approach for quantifying system cost uncertainty; and one which offers a set of mechanisms for treating the important technical issues described in this paper. Furthermore, the analytic nature of the ACOP model facilitates its implementation on any electronic spreadsheet, and therefore, lessens the required execution time in comparison to a simulation-based approach.

1.0 Introduction

Cost uncertainty analysis quantifies known technical and acquisition risks of a system to provide a probability distribution around the estimated cost (e.g., an interval in which the true cost hopefully resides), and an assessment of the level of reserve funds to recommend for the program. Cost uncertainty analysis also identifies elements of a system that are associated with significant amounts of cost risk. This fosters early recognition of inadequately specified requirements and permits risk mitigating decisions to be made early in the system definition phases. For the program manager, cost uncertainty analysis reveals the system cost impact of uncertainty in such areas as: technical realism, the soundness of the acquisition strategy, the reasonableness of the schedule, and the accuracy of the estimation methodology. These are among the areas of risk that collectively contribute to determining the level of confidence in a system's cost estimate.

In 1987, The MITRE Cost Analysis Technical Center derived a planning estimate of a system consisting of a sophisticated set of technical objectives to be achieved in a highly complex and aggressive acquisition approach. It was necessary to develop a cost analysis methodology robust enough to be sensitive to the numerous technical uncertainties driving the cost of the acquisition. In addition to identifying the technical risks, the analysis was required to consider the numerous interdependencies existing between the engineering activities, project milestones, end-item deliverables, and the acquisition approach of multiple segments with multiple subcontracts embedded in four overlapping releases. The technical nature and acquisition approach of this system was so complex, that our set of simulation models available at that time could not adequately quantify the uncertainties described above. Furthermore, the cost spreadsheet model for this system (Excel Version 1.5) consumed nearly one megabyte of storage, so even if a simulation front-end had been created to perform a classic Monte Carlo analysis, the amount of execution time and resources required to run the simulation would have been excessive. Therefore, a general analytic approach was developed to model the complex character of this system while also operating within available computational resources. This paper presents the Analytic Cost Probability model which is sufficiently general to be applicable to a wide spectrum of programs.

2.0 Foundations Of The Analytic Cost Probability Model

The matrix shown in table 2-1 depicts cost elements of a generalized work breakdown structure (WBS) where CE_i represents the ith element in the system. The random vectors $\mathbf{Y_1}$, $\mathbf{Y_2}, \ldots, \mathbf{Y_n}$ represent n-categories of cost uncertainty identified for the system. Among the categories which these distributions might represent are: risks inherent to the technical approach, the soundness of the acquisition strategy, the reasonableness of the schedule, and the accuracy of the estimation methodology. The general structure provided by the random vectors $\mathbf{Y_1}, \mathbf{Y_2}, \ldots, \mathbf{Y_n}$ enables the analyst to separate or combine categories of cost uncertainty (e.g., uncertainty in labor or overhead rates can either be described by a unique random vector, or in combination with a random vector representing the overall cost uncertainty) unique to the system.

From the matrix shown in table 2-1 the total system cost is defined by

$$\text{Cost} = \sum\sum Y_{ij} \; (j = 1, ..., n), \; (i = 1, ..., m) \qquad (2\text{-}1)$$

where Y_{ij} is a set of random variables independent in j and correlated in i. That is, the columns denoted by the index j represent independent (uncorrelated) categories of cost uncertainty, and the rows denoted by the index i allow for correlation to exist between the cost elements in the WBS.

Table 2-1

A Generalized System Work Breakdown Structure

Cost Element **CE**	$\mathbf{Y_1}$	• • •	$\mathbf{Y_n}$
CE_1	Y_{11}	• • •	Y_{1n}
CE_2	Y_{21}	• • •	Y_{2n}
•	•		•
•	•		•
•	•		•
CE_m	Y_{m1}	• • •	Y_{mn}

Each Y_{ij}, in table 2-1, represents an assumed probability distribution around the estimated cost of CE_i (e.g., an interval in which the true cost hopefully resides) according to the specific cost uncertainty category. Specifically, the random variables Y_{i1} are described by probability distributions whose modes represent the estimated cost of each element. This cost is referred to as the *point estimate*. The point estimate is the cost derived for an element in the WBS without including the effect of uncertainties in the element's technical definition or cost estimation methodology. The random variables Y_{in}, for $n \geq 2$, are described by probability distributions with modes set equal to zero and whose ranges reflect a positive and/or negative cost delta around the point estimate, given by the mode of Y_{i1}.

General mathematical expressions for the overall system cost mean and variance, denoted by E(Cost) and Var(Cost), respectively, are given by:

$$E(\text{Cost}) = E(\sum\sum Y_{ij}) = \sum\sum E(Y_{ij}) \tag{2-2}$$

and

$$\text{Var}(\text{Cost}) = \text{Var}(\sum\sum Y_{ij}) = \sum \text{Var}(\sum Y_{ij}) \tag{2-3}$$

where

$$\text{Var}(\sum Y_{ij}) = \sum \text{Var}(Y_{ij}) + 2\sum \text{Cov}(Y_{ij}, Y_{kj}) \text{ for } k > i \tag{2-4}$$

The covariance between any two cost elements is given by

$$\text{Cov}(Y_{ij}, Y_{kj}) = \rho_{ikj}\, \sigma_{ij}\, \sigma_{kj} \tag{2-5}$$

where σ_{ij} and σ_{kj} are the standard deviations of Y_{ij} and Y_{kj}, respectively, and ρ_{ikj} is the coefficient of correlation between the cost elements Y_{ij} and Y_{kj}. A full treatment of cost element correlation will be presented in section 3.

It can be seen from the above equations that correlated random variables affect only the magnitude of the system's cost variance and not its mean. Although the generalized form of the system cost variance given by equation 2-3 is valid for all instances, the following will discuss how the shape of the system cost probability distribution is directly affected by whether correlation exists between the cost elements in the system.

2.1 Formulating The System Cost Probability Distribution

This section provides assumptions under which the system cost probability distribution may be considered approximately normal. The following will present a discussion of these assumptions and how they contribute towards the analytical formulation of the system cost probability distribution.

When all cost elements of a system are assumed to be independent, the classical Central Limit Theorem [1] can be invoked to assert that the system cost probability distribution is approximately normal, with mean E(Cost) given by equation 2-2, and variance Var(Cost) given below by equation 2-6.

$$\mathrm{Var(Cost)} = \sum\sum \mathrm{Var}(Y_{ij}) \qquad (2\text{-}6)$$

However, the assumption of independence between all the elements of a system is, in most cases, an oversimplification of the manner in which the engineering activities defined by these elements tend to behave. In cases where cost elements tend to exhibit positive correlation, an underestimate of the system cost variance is likely under the assumption of independence since independent random variables have zero correlation, that is, the covariance contribution to Var(Cost), given by equation 2-3, is zero. Clearly, most systems exhibit varying degrees of correlation between the engineering activities defined by their representative cost elements; and a further discussion of this will be presented in section 3. The following focuses on where cost element correlation is likely to occur in a work breakdown structure, and arguments are presented for describing the shape of the system cost probability distribution when correlation between cost elements have been identified to exist in the system.

Elements of a system which drive its cost are essentially those which represent the hardware and software architecture developed to meet the requirements of the specification. In the terminology of MIL-STD-881A, these elements combine in cost to a single element known as the prime mission product (PMP). It is, therefore, in the PMP cost element where the greatest technical and acquisition risks of a system typically exist. Thus, it is not unusual for the PMP to exhibit the dominant cost variance in the system when compared to the cost variances of the other cost elements, such as, system engineering, program management, and system test and evaluation. In theory, the engineering activities defined by these elements, which represent contract support, should be highly correlated to the system's PMP. If the following conditions exist among the cost elements in the system:

- the PMP has dominant cost variance and,
- the contract support cost elements are highly correlated to PMP,

then it is reasonable to assert that the total system cost probability distribution is approximated by the distribution of PMP with mean given by equation 2-2 and variance given by equation 2-3.

In many systems the shape of the PMP distribution may be considered normal. This can occur in software intensive systems (an increasing trend in major ESD acquisitions) where the shape of the PMP distribution is essentially governed by the shape of the software distribution which, under certain general conditions [2], has been shown to be asymptotically normal. In circumstances where normality does not describe the shape of the PMP distribution (e.g., if the PMP distribution is skewed) an approximation of the total system cost probability distribution through formal simulation techniques may be warranted.

It is important to remember that the purpose of cost uncertainty analysis is to provide the program manager with a cost that has an acceptable probability of being exceeded. Acquisitions will always face the unknown unknowns which cannot be quantified by this analysis, and therefore, the confidence in the cost and the shape of the system cost probability distribution should never be construed as anything more than an approximation.

3.0 Cost Element Correlation

It is typically an oversimplification to assume that all of the engineering activities defined by the cost elements of a system are independent. Such assumptions can yield low estimates of a system's cost variance, and therefore, unrealistically small differences in cost between the tails of the probability distribution is likely to be seen. In a work breakdown structure consisting of m cost elements the variance can be underestimated by up to a factor of m, if that variance was determined from the assumption of independence between the engineering activities in the system. Clearly, correlation among the elements of cost can be an important contributor to the magnitude of the system cost variance.

The correlation coefficient ρ_{ikj} is used to measure of the level of interdependence existing between the pairs of engineering activities represented by the cost elements in a system. Ideally, the correlation coefficient should be derived, if historical cost data are available in sufficient quantity and completeness. Recently, we derived from the USAF Electronic Systems Division (ESD) cost data base, the sample correlation coefficient r between selected pairs of MIL-STD-881A cost elements. Table 3-1 presents a view into the correlations derived from this data base, when the contract support cost elements were paired with the system's PMP cost element. Table 3-1 also presents the coefficient of determination r^2. The coefficient of determination gives the proportion of the total variability of the cost for the contract support elements that can be accounted for by the PMP cost element.

Table 3-1

Sample Correlation Coefficients - ESD Cost Data Base

Contract Support Elements	Sample Correlation r With PMP Cost Element	r^2
System Engineering	0.82	0.67
Program Management	0.82	0.67
System Test & Evaluation	0.73	0.53
Data	0.70	0.49
Operational Site Activation	0.57	0.32
Training	0.57	0.32

In circumstances where it is not possible or practical to collect enough historical data to derive the sample correlation coefficients, table 3-2 was developed as a basis for subjectively assigning specific values of ρ_{ikj} to each cost element pair. The values in table 3-2 are only suggested and, if necessary, should be amended to reflect experiences on past programs specific to an organization.

Since the number of pairs of elements which define the engineering activities of a system can be quite large, it may be desirable to bound the effect of correlation on the system cost variance, rather than investing the effort to either derive or specify the correlation for each element pair explicitly. Appendix A presents a method for bounding the effect that pairwise correlation between cost elements can have on the system cost variance.

Table 3-2

Suggested Values Of ρ_{ikj}

ρ_{ikj}	Description
0	There is **no** correlation between the pair of cost elements given by Y_{ij} and Y_{kj}
0.25	There is **some** correlation between the pair of cost elements given by Y_{ij} and Y_{kj}
0.50	There is **moderate** correlation between the pair of cost elements given by Y_{ij} and Y_{kj}
0.75	There is **strong** correlation between the pair of cost elements given by Y_{ij} and Y_{kj}
1	There is **complete** correlation between the pair of cost elements given by Y_{ij} and Y_{kj}

4.0 Implementing The ACOP Equations - An Illustration

A simple illustration on the use of the fundamental equations developed for the ACOP model is presented in this section. Table 4-1 provides a hypothetical system cost estimate shown by the elements of its work breakdown structure. Two categories of cost uncertainty $\mathbf{Y_1}$ and $\mathbf{Y_2}$ are defined in table 4-1. It is assumed that $\mathbf{Y_1}$ represents estimation uncertainty and $\mathbf{Y_2}$ represents cost uncertainty due to probable variations in the acquisition schedule. Estimation uncertainty is defined to include any one or combination of the following: errors inherent with cost models, costing techniques, and labor rates; and areas of the technical baseline that may be inadequately specified. The illustration starts with the distributions shown for the elements as already determined from a technical assessment of the risks perceived to exist in each cost element.

Table 4-1

A Hypothetical System Cost Estimate*
(Dollars Million)

Cost Element CE	Y_1	Y_2
Prime Mission Product (PMP)		
	Y_{11}~N(12.5, 6.6)	Y_{12} ~ Trng(-2, 0, 3)
Contract Support Elements		
System Engineering and PManagement	Y_{21}~ Trng(4.0, 7.6, 11.4)	Y_{22} ~ Trng(-1, 0, 2)
System Test & Evaluation	Y_{31}~ Trng(3.1, 4.5, 9.7)	Y_{32} ~ Trng(-1.5, 0, 3)
Data	Y_{41}~ Unif(0.9, 1.2)	$Y_{42} = 0$
Site Activation	Y_{51}~ Trng(5.1, 6.6, 12.1)	Y_{52} ~ Trng(-1, 0, 3)
Spares	Y_{61}~ Unif(1.0, 1.5)	$Y_{62} = 0$
Warranty	Y_{71}~ Unif(0.9, 1.3)	$Y_{72} = 0$
Prototype	Y_{81}~ Trng(1.0, 1.5, 2.4)	$Y_{82} = 0$
Computer Operations Support	Y_{91}~ Trng(0.9, 1.2, 1.6)	$Y_{92} = 0$
Training	Y_{101}~ N(3.3, 1.0)	$Y_{102} = 0$

Note: * N = Normal Distribution, Trng = Triangular Distribution, Unif = Uniform Distribution

From equation 2-1 the total system cost is defined by:

$$\text{Cost} = \sum\sum Y_{ij} \qquad j = 1, 2 \text{ and } i = 1, \ldots, m = 10$$

Cost summary statistics for the hypothetical system described in table 4-1 are presented in table 4-2. Formulas for the mean and variance of each Y_{ij} shown in table 4-1 may be found in reference one.

Table 4-2

Cost Summary Statistics

Cost Element CE	$E(Y_{i1})$ (\$M)	$E(Y_{i2})$ (\$M)	$Var(Y_{i1})$ $(\$M)^2$	$Var(Y_{i2})$ $(\$M)^2$
Prime Mission Product (PMP)				
	12.5	0.33	6.6	1.06
Contract Support Elements				
System Engineering and PManagement	7.7	0.33	2.2	0.39
System Test & Evaluation	5.8	0.50	2.0	0.88
Data	1.0	0	0.01	0
Site Activation	7.9	0.67	2.3	0.72
Spares	1.3	0	0.02	0
Warranty	1.1	0	0.01	0
Prototype	1.6	0	0.08	0
Computer Operations Support	1.2	0	0.02	0
Training	3.3	0	1.0	0
Total	43.4	1.8	14.2	3.05

From equation 2-2 the total mean cost of the system is given by:

$$E(\text{Cost}) = E(\sum\sum Y_{ij}) = \sum\sum E(Y_{ij}) = 45.2$$

and the total cost variance of the system, given by equation 2-3, is:

$$\begin{aligned} \text{Var}(\text{Cost}) &= \text{Var}(\sum\sum Y_{ij}) = \sum \text{Var}(\sum Y_{ij}) \\ &= \text{Var}(\sum Y_{i1}) + \text{Var}(\sum Y_{i2}) \end{aligned}$$

where j = 1, 2 in this illustration. From equation 2-4 we have:

$$\text{Var}(\sum Y_{i1}) = \sum \text{Var}(Y_{i1}) + 2\sum \rho_{ik1}\sigma_{i1}\sigma_{k1} \quad (i < k)$$

and

$$\text{Var}(\sum Y_{i2}) = \sum \text{Var}(Y_{i2}) + 2\sum \rho_{ik2}\sigma_{i2}\sigma_{k2} \quad (i < k)$$

The terms σ_{ij} and σ_{kj} denote the standard deviations of Y_{ij} and Y_{kj}, respectively, and ρ_{ikj} is the coefficient of correlation between the cost elements Y_{ij} and Y_{kj}. Suppose an analysis of this system revealed the existence of a strong level of correlation ($\rho = 0.75$) between the system engineering through spares cost elements with the PMP element. Then, with h = 2, . . ., 6 the total system cost variance is:

$$\begin{aligned} \text{Var}(\text{Cost}) &= [\sum \text{Var}(Y_{i1}) + 2(0.75)\,\sigma_{11}\sum \sigma_{h1}] \\ &\quad + [\sum \text{Var}(Y_{i2}) + 2(0.75)\,\sigma_{12}\sum \sigma_{h2}] = 38.9 \end{aligned}$$

Since the PMP cost element has the dominant cost variance in this system, and the contract support cost elements are strongly correlated to PMP, it may asserted that the distribution of total cost is approximated by the distribution of the PMP, which in this case is given to be normal. Table 4-3 presents the cost probability distribution for this system based on a normal with mean 45.2 and variance 38.9.

Table 4-3

Cumulative Probability Distribution of System Cost

Probability (Percentile)	Cost ($M)
5	35.0
10	37.2
50	45.2
90	53.2
95	55.4

The information obtained from carefully quantifying the uncertainties inherent to this system yielded not only an estimate of the cost that is equally likely to overrun or underrun ($45.2M), but also the amount of cost reserve needed to obtain high confidence in the final recommended estimate. For instance, a reserve of $8M added to the mean brings the program's estimate to the 90 percentile confidence level. Table 4-3, therefore, provides insights to the decision maker on the amount of cost reserve likely to be required, to ensure budgeting the program at an acceptable level of confidence.

5.0 Summary Remarks

Although various cost uncertainty analysis methods have been available to the industry for many years, among the contributions to the field provided by the ACOP model are the following:

- the general structure given by the random vectors $\mathbf{Y}_1, \mathbf{Y}_2, \ldots, \mathbf{Y}_n$ enables the analyst to identify n-categories of cost uncertainty and isolate the effect of any one category on the total cost risk;

- the methods for treating pairwise correlation between the cost elements of the system, and identifying when a system's cumulative cost probability distribution may be approximated by the normal, even when all the elements of cost are *not* independent;

- the development of a useful theorem (presented in appendices A and B) for bounding the effect of correlation on the system cost variance.

Thus, we now have a fully developed analytic approach for quantifying system cost uncertainty; and one which offers a set of mechanisms for treating the important technical issues described in this paper. Furthermore, the analytic nature of the ACOP model facilitates its implementation on any electronic spreadsheet, and therefore, lessens the required execution time in comparison to a simulation-based approach.

The paper cited in reference two was recently published by the IEEE Computer Society Press as part of a collection of technical papers on software risk management. That paper provided methods (analytic and simulation) developed for determining the software cost probability distribution for estimates generated from αI^{β} models. The paper presented herein, has expanded on the concepts developed in reference two and provides an analytic approach for the more general problem of system cost uncertainty. Together, these two papers provide the cost community with a complete set of uncertainty analysis methods that have recently evolved at MITRE, to support increasingly more complex cost and cost risk studies for the Electronic Systems Division.

References

1. Bickel P. J., Doksum K. A., Mathematical Statistics-Basic Ideas and Selected Topics, Holden-Day, San Francisco, California, 1977.

2. Garvey P. R., Powell F. D., "Three Methods For Quantifying Software Development Effort Uncertainty," published in Software Risk Management, Boehm B. W., IEEE Computer Society Press, November 1989.

3. Taylor A. E., Mann W. R., Advanced Calculus, Xerox College Publishing, Lexington, Massachusetts, 1972

Appendix A

Bounding The Effects of Cost Element Correlation In A System Work Breakdown Structure

There can be an imposing implementation problem associated with the approach presented in section 3. Since a system consisting of *m* cost elements contains exactly $m(m-1)/2$ distinct cost element pairs, each pair of elements requires $m(m-1)/2$ specific determinations of ρ_{ij}. In a work breakdown structure consisting of 20 cost elements, an examination of the correlation between 190 distinct element pairs would be required. Clearly, there are circumstances when examining correlation at the element level can be impractical.

Therefore, an alternative to the approach presented in section 3 has been developed to minimize the number of correlation assessments required, without losing a means of capturing the potential effect that pairwise correlation between cost elements can have on the total system cost variance. It is based on an examination of the theoretical bounds on the system cost variance due to the presence of correlation between the cost element pairs in a work breakdown structure.

A.1 Cost Element Correlation - Theoretical Bounds

This section presents a theorem which establishes the theoretical bounds on the total system variance due to the existence of correlation in a system work breakdown structure. A proof of the theorem is offered in Appendix B. A method is then formulated around this theorem for bounding the effects of correlation on the system cost variance.

Theorem 1.

Suppose we let S be the sum of m cost element random variables $X_1, X_2, \ldots, X_m$ each having finite variance. Define σ^2_S and σ^2_0 by

$$\mathrm{Var}(S) = \sigma_s^2 = \sum_{i=1}^{m} \sigma^2_{X_i} + 2\sum_{i<j} \rho_{ij} \sigma_{X_i} \sigma_{X_j}$$

and

$$\sigma_0^2 = \sum_{i=1}^{m} \sigma_{X_i}^2$$

If $\rho = \max(\rho_{ij}, 0)$ then Var(S) is bounded by:

$$\sigma^2_S \leq [1 + \rho(m - 1)]\, \sigma^2_0 \qquad \text{(A-1)}$$

From a cost perspective, the inequality presented by A-1 is the bound on the system variance due to the presence of correlation in the system work breakdown structure. For convenience, inequality A-1 will be referred to as the correlation bound.

In its present form Theorem 1 yields very conservative correlation bounds even in systems with a typical number of cost elements. For instance, in a system consisting of 20 cost elements with $\rho = 0.75$, the correlation bound is approximately:

$$\sigma^2_S \leq 15\sigma^2_0$$

Clearly, this overstates the impact since not all the elements in a system work breakdown structure are likely to be pairwise correlated with a correlation coefficient of 0.75. There typically exists a mixture of correlated and uncorrelated elements in a system. Therefore, Theorem 1 is too conservative for bounding the effect of correlation on the total system cost variance when applied on an element-by-element basis. The following will develop an approach which reduces the *m*-dimensional aspects of the problem into a set of five distinct cost element random variable groups. Bounding the total system cost variance σ^2_S due to the effects of correlation will be determined by bounding the variance of each group using Theorem 1.

Consider a system which can be aggregated into groups defined to be:

Z_1 = Prime Mission Product
Z_2 = System Engineering/ Program Management
Z_3 = MIL-STD-881A Contract Support Elements (Excluding Z_2)
Z_4 = Unique Contract Support Elements (Excluding Z_2 and Z_3)
Z_5 = System Program Office Costs

Each group contains the cost elements X_{kv} ($v = 1, \ldots, n_k$) and n_k denotes the number of cost elements in group Z_k. Let $\sigma^2_{0,g}$ represent the sum of the group variances given by:

$$\sigma^2_{0,g} = \sum_{k=1}^{5} \sigma^2_{Z_k}$$

where

$$\sigma^2_{Z_k} = \sum_{v=1}^{n_k} \sigma^2_{X_{kv}} + 2\sum_{v<h} \rho_{kv,kh}\, \sigma_{X_{kv}}\, \sigma_{X_{kh}}$$

Define

$$\sigma^2_0 = \sum_{k=1}^{5} \sigma^2_{0,k}$$

where

$$\sigma^2_{0,k} = \sum_{v=1}^{n_k} \sigma^2_{X_{kv}}$$

When Theorem 1 is applied to each group we have for all k

$$\sigma^2_{Z_k} \leq [1 + \rho'_k(n_k - 1)]\,\sigma^2_{0,k}$$

where $\rho'_k = \max[(\rho_{kv,kh}), 0]$ in group k; therefore

$$\sigma^2_{0,g} = \sum_{k=1}^{5} \sigma^2_{Z_k} \leq \sum_{k=1}^{5} [1 + \rho'_k(n_k - 1)]\,\sigma^2_{0,k}$$

Define ζ to be

$$\zeta = \frac{\sum_{k=1}^{5} [1 + \rho'_k(n_k - 1)]\,\sigma^2_{0,k}}{\sigma^2_0}$$

where ζ can be determined for each system since n_k is known, ρ'_k can be assessed, and $\sigma^2_{0,k}$ and σ^2_0 are computed from the cost element probability density functions. Next, we will use ζ and Theorem 1 to bound σ^2_s.

If $\rho_g \geq 0$ is the maximum correlation coefficient between the ten pairs of cost element random variable groups stated above, then applying Theorem 1 to the group aggregations yields:

$$\sigma^2_s \leq (1 + 4\rho_g)\sigma^2_{0,g} \tag{A-2}$$
$$\sigma^2_s \leq (1 + 4\rho_g)\zeta\sigma^2_0 \tag{A-3}$$
$$\sigma^2_s \leq 5\zeta\sigma^2_0 \quad \text{if } \rho_g = 1 \tag{A-4}$$

where σ^2_s and σ^2_0 are as defined in Theorem 1.

In summary, the methodology developed in this appendix provides a technique for bounding the effects of cost element correlation on the estimated system cost variance. The main feature of this approach is the elimination of the need to examine correlation on an element-by-element basis. When implementing this technique first determine ζ for the system, then for a specific ρ_g, equation A-3 provides a bound on the total system variance σ^2_s. A formal proof of the theorem stated in this appendix is offered in Appendix B.

Appendix B

Theorem

Let S be the sum of n random variables $Y_1, Y_2, \ldots, Y_n$ each having finite variance. Given

$$\sigma_S^2 = \sum_{i=1}^{n} \sigma_{Y_i}^2 + 2\sum_{i<j} \rho_{ij} \sigma_{Y_i} \sigma_{Y_j} \tag{B-1}$$

let $\rho = \max(\rho_{ij}, 0)$ where ρ_{ij} is the correlation between Y_i and Y_j; then

$$\frac{\sigma_S^2}{\sigma_0^2} \leq 1 + \rho(n-1)$$

where

$$\sigma_0^2 = \sum_{i=1}^{m} \sigma_{Y_i}^2$$

Proof:

The approach is to show that σ^2_S/σ^2_0 is at a maximum when $\sigma_i = \sigma_j$ for all i and j; and then to evaluate σ^2_S/σ^2_0 when $\sigma_i = \sigma_j$ for all i, j. For notational convenience let:

$$\sigma_{Y_i} = \sigma_i$$

From equation (B-1) we have

$$\frac{\sigma_S^2}{\sigma_0^2} = \frac{\sum_{i=1}^{n} \sigma_i^2 + 2\sum_{i<j} \rho_{ij} \sigma_i \sigma_j}{\sum_{i=1}^{n} \sigma_i^2} \leq \frac{\sum_{i=1}^{n} \sigma_i^2 + 2\rho\sum_{i<j} \sigma_i \sigma_j}{\sum_{i=1}^{n} \sigma_i^2} = 1 + \frac{2\rho\sum_{i<j} \sigma_i \sigma_j}{\sum_{i=1}^{n} \sigma_i^2} \tag{B-2}$$

For any constant c given that

$$\sum_{i=1}^{n} \sigma_i^2 = c$$

the maximum of equation B-2 occurs when

$$2\rho \sum_{i<j} \sigma_i \sigma_j$$

is maximized subject to the constraint:

$$\sum_{i=1}^{n} \sigma_i^2 = c$$

The method of Lagrange [3] will be used to find this maximum. Form the function:

$$U = 2\rho \sum_{i<j} \sigma_i \sigma_j + \lambda[\sum_{i=1}^{n} \sigma_i^2 - c]$$

from which

$$\frac{\partial U}{\partial \sigma_i} = 2\rho \sum_{\substack{k=1 \\ i \neq k}}^{n} \sigma_k + 2\lambda \sigma_i = 0$$

thus

$$\lambda = \frac{-\left(\rho \sum_{\substack{i=1 \\ i \neq k}}^{n} \sigma_k \right)}{\sigma_i}$$

Therefore, for i = 1 and any j = 2, 3, . . ., n we have

$$\sigma_j \sum_{k=2}^{n} \sigma_k = \sigma_1 \sum_{\substack{k=1 \\ j \neq k}}^{n} \sigma_k$$

The above equation can be factored into:

$$\sigma_j\left\{\sigma_j + \sum_{\substack{k=2\\k\neq j}}^{n}\sigma_k\right\} = \sigma_1\left\{\sigma_1 + \sum_{\substack{k=2\\k\neq j}}^{n}\sigma_k\right\}$$

from which it follows that $\sigma_1 = \sigma_j$. We therefore have a maximum for $\sigma^2{}_S/\sigma^2{}_0$ when $\sigma Y_1 = \sigma Y_j = \sigma$. It then follows from equation (B-2) that

$$\frac{\sigma_S^2}{\sigma_0^2} \leq 1 + 2\rho\frac{[(n-1)n]\sigma^2/2}{n\sigma^2}$$

from which

$$\frac{\sigma_S^2}{\sigma_0^2} \leq 1 + \rho(n-1)$$

THE INCORPORATION OF UNCERTAINTY INTO INVESTMENT EVALUATIONS

by

William Resnick

Department of Chemical Engineering
Israel Institute of Technology
Haifa, Israel

ABSTRACT

Evaluations of investment proposals based on forecast cash flows are subject to uncertainties and the more remote in time a cash flow is expected to occur the greater the degree of uncertainty. A simple technique is proposed here which manipulates the discounting procedures so that late-occurring cash flows are penalized more severely in present value terms than they would be by customary discounting. The technique proposed transforms calendar time to risk-time by means of a risk-exponent but otherwise uses conventional profitability measures. Guidance in the choice of the appropriate values for the risk-exponent is given.

INTRODUCTION

Evaluations of the economic merit of investment proposals are based on cash flows which, in reality, are forecasts and projections into the future. The profitability analyses which result are produced, therefore, against a backdrop of uncertainties in future cash flows and uncertain time events. The evaluators who are performing the economic evaluations which are based on these forecasts and projections are, in a manner of speaking, attempting to prophecy the future.

As has been noted some time ago by Rabbi Eudemus [1] "the gift of prophecy has been denied to prophets and bestowed upon scholars" although this view has been challenged by his colleague, Rabbi Johanan [2], who claimed that the gift of prophecy was bestowed "upon fools and children". Although hoping that the Eudemus theory is the correct one, those scholars who deal with investment appraisals and profitability analyses involving the future do not relish the possibility that the Johanan theory might prove to be valid. Recognizing that uncertainty exists, they have attempted to quantify this uncertainty so as to be able to deal with it in investment proposals.

One procedure for dealing with uncertainty is to adjust the corporate profitability criteria so that an investment proposal with a high degree of risk would be required to meet more stringent criteria than would a proposal with a lower degree of risk or uncertainty. The usual practice is to use a risk-adjusted discount rate in the evaluation of the high-risk project which is higher than that which is used as the interest rate for low-risk projects. English [3] shows, however, that the risk-rate-of- return instead of being increased as compared to the low risk case should, under certain circumstances, be reduced. He demonstrates that this is true, especially, when future investments are projected.

Several authors [4,5,6] have developed functional forms for risk-adjusted discount rates. Some of the major drawbacks were noted by Treynor and Black [5] who pointed out that it was difficult to use their value equations because of the difficulty of estimating cash flows as a function of economic variables. They conclude that the greatest value of the equations probably would be in theoretical investigations and as a background for thinking about investment decisions. As Bhattacharya [6] states the procedures and equations developed are relatively complex and attempts to simplify them result in the loss of theoretical rigor. The development of tractable equations could prove useful in the area of capital budgeting under uncertainty.

A number of authors have dealt with risk by modelling cash flows and developing probability functions on the basis of certain assumptions and usually for relatively specific applications for cases in which cash flows and/or their timings are uncertain[7-13]. Steece [11] pointed out in his ex-ante study of measures for evaluating forecast models that forecasts at different lead times are correlated with each other but the standard errors of forecasts increase with lead time.

The various models and probability functions developed are usually based on the assumption that a pattern, trend or behavior which developed in the past will continue on into the future. In most cases, however, no historical precedent would be available for establishing trends and patterns for the majority of elements which enter into the cash flow calculations. This would obviously be the case when dealing, for example, with an investment proposal involving a new product or process. In addition, competitors entering a market could make nonsense of any probability-based evaluation.

Other approaches involve expert opinion in which, generally, three-point probability functions are derived from opinions as to the most-likely, lowest-likely and highest-likely values of the cash flow elements. These values can then be used to produce "subjective" probabilities to quantify the predictions made by expert opinion concerning a future event.

Although the use of probability functions permits the uncertainty problem to be "solved", it should be apparent from the above that it is solved by the use of arbitrary assumptions which result in some mathematical expression. This expression, in turn, can then be subjected to a rigorous and academically-satisfying mathematical treatment.

Dean [14] pointed out some time ago that adjustments for uncertainty may be challenged as being nothing more than guesses and that probability adjustments merely disguise and transform the guess element. The disguised form than reappears in some other stage of the decision-making process.

The procedures presently available for dealing with uncertainty either require that the decision-makers apply and use different criteria than those to which they are accustomed or else that they resort to complicated and difficult methodologies. The results of the analyses described above are usually presented to the decision-maker as net present worths or as internal rates of return which are expressed as probabilities. It has been mentioned [15] that this presentation is not particularly welcomed by most decision-makers who must now choose between the possibilities of disaster or spectacular success.

In this paper a simple procedure is proposed involving time-dependent risk-adjusted discounting rates. These time-dependent risk-adjusted rates meet the objections raised by English [3].

In the methodology proposed here conventional profitability measures and indices are still used for profitability evaluation but the uncertainty elements are incorporated into the conventional profitability measure as one of its inherent elements. The decision-maker can use this proposed procedure and continue to apply, not only the same criteria which he or she had been accustomed to using previously, but also the same numerical values for these criteria.

ANALYSIS OF INVESTMENT PROPOSALS

The forecasting of capital requirements requires that the cost of equipment, the cost of its erection and installation, construction time and start-up costs be estimated while the estimates of cash inflow require that markets, selling prices and operating costs be predicted. Uncertainties in the forecasting of markets, prices and sales are usually much greater than those involved in the estimation of capital requirements for a plant. The result is that the capital requirements for a project can generally be predicted with reasonable accuracy' but the expected revenues are subject to many more uncertainties in elements such as sales volume, selling price, cost of raw materials, and the cost of labor and utilities. In addition, capital outflow generally occurs over a relatively short period of time but the revenues generated from product sales which are received subsequent to the capital investment flow over a long period of time. Furthermore, the more remote in time the cash flow is forecast to occur, the greater will be the degree of uncertainty in the forecast cash flows.

Any technique which is to be used to incorporate uncertainty considerations into project appraisals must be tailored so as to take into account the uncertainty factors mentioned above and to deal with them in an appropriate manner. Sinclair [16] has already mentioned that weighting factors other than those applied in conventional discounting techniques can be used and that it is possible to tailor discount factors in accordance with the effect which is to be achieved. A suitable technique for dealing with uncertainties with a cash-flow pattern as described above would be one which would manipulate the discount factors so that cash flows expected to occur in the more distant future would be discounted more severely than they would be by conventional interest compounding discounting. In other words the risk adjustment would be applied to time instead of to the interest rate.

INCORPORATION OF UNCERTAINTY CONSIDERATIONS

A discounting technique which can take into account the uncertainty pattern discussed above and deal with it in an appropriate manner would be to weight time, or the number of discounting periods, with a risk-exponent, r, by the use of t^r rather than t in the various discounting expressions. The result is a time-dependent risk-adjusted discount factor and the term tr can be considered to be a "risk-time". This risk-time can be used in place of time in the usual discounting expressions. Thus, the discount factor for an instantaneous cash flow occurring at time t would be $1/(1+i)^\wedge t^r$ for the case of discrete interest compounding and $\exp(-it^r)$for continuous interest compounding. The interest rate, i, to be used will be the usual hurdle interest rate whereas the risk element is now imbedded in the risk-time, t^r, by virtue of the risk-exponent, r.

The present worth of a future cash flow would be

$$P=F/(1+i)^{t^r}$$

for discrete interest compounding and

$$P=F*(\exp(-it^r))$$

for continuous interest compounding. The Net Present Worth of a series of cash flows would, of course, be the sum of the individual net worths. Any of the usual expressions for the time value of money can be used by replacing the time term with the risk-time, t^r.

Only when the value of the risk-exponent is taken as unity would the discounting be in accordance with the usual interest compounding where the time value of money is determined only by the hurdle rate. For values of r greater than unity, however, discounting is greater than that attributable only to the time value of money at the hurdle rate. Thus, revenues expected in the distant future, and these are precisely the revenues which are subject to the greatest uncertainties, would also be much more heavily discounted than would revenues which are expected to occur sooner.

On the other hand, values of r which are smaller than unity would discount a future payment at a lower rate than that attributable to usual interest compounding. Alternatively, the future worth of a present sum would be larger than that attributable to usual interest compounding. This is precisely the requirement one would need to deal with uncertainties in the projected capital requirements and would amplify the uncertainties inherent in investment estimates projected to occur in the more-distant future. Thus, the risk-exponent tailors the discounting behavior in accordance with the effect to be achieved [18].

DEMONSTRATION OF EFFECT OF RISK-EXPONENT

The effect of using the risk-exponent on discount factors can be observed in Table 1. Values of the risk-exponent weighted discount factors and the equivalent "risk" interest rate are shown for the hurdle rate of 15% at risk-exponent values of 0.65, 1.00 and 1.20 for the case of discrete interest compounding. It is apparent that cash flows are discounted more heavily for values of r greater than unity, i.e., when the risk-time is greater than the calendar time. Thus, at r = 1.2 the discount factor to calculate the present worth of a cash flow projected to occur 10 years in the future is equal to 0.1091 (equivalent to an interest rate of 24.8%). This is to be compared to 0.2472 when no time-risk dependency is assumed, i.e., when the value of the risk-exponent is taken as unity and the risk-time and calendar time are equal. It should be noted that the discount factor and equivalent interest rate are independent of the value of the risk exponent for cash flows that occur at time zero or at year one and depend only on the nominal interest rate.

It can also be seen from the same table that a projected capital expenditure in, say, three years would be discounted to present value by a factor of 0.6575 if r is taken equal to unity. If, however, when risk considerations on the estimated investment are included in the form of a risk-exponent equal to 0.65, the sum would be discounted by a factor of 0.7517.

Table 1

Discount Factors and Risk Interest Rates
for Several Values of Risk Exponent
At 15% Interest Rate

Time Years	Risk Exponent					
	0.65		1.0		1.2	
	Discount Factor	Equivalent Interest Rate, %	Discount Factor	Equivalent Interest Rate, %	Discount Factor	Equivalent Interest Rate, %
0	1.000	15.0	1.0000	15.0	1.0000	15.0
1	0.8696	15.0	0.8696	15.0	0.8696	15.0
2	0.8031	11.6	0.7561	15.0	0.7254	17.4
3	0.7517	10.0	0.6575	15.0	0.5931	19.0
4	0.7088	9.0	0.5718	15.0	0.4782	20.3
5	0.6718	8.3	0.4972	15.0	0.3813	21.3
6	0.6390	7.8	0.4323	15.0	0.3012	22.1
7	0.6095	7.3	0.3759	15.0	0.2360	22.9
8	0.5827	7.0	0.3269	15.0	0.1837	23.6
9	0.5582	6.7	0.2843	15.0	0.1420	24.2
10	0.5356	6.4	0.2472	15.0	0.1091	24.8

Thus, if $r > 1$ is used as the value for the risk-exponent in the discounting of cash revenues and $r < 1$ is used to discount capital investment, the risk-exponent can be considered to amplify uncertainties in a pessimistic direction. It would attenuate the present value of positive cash flows and would amplify the present value of investments which are projected to occur in the future. This treatment of uncertainty is consistent with the risk-averse nature of most managers and decision-makers.

EVALUATION OF AN EXAMPLE INVESTMENT PROPOSAL

The cash flows for the example capital project to be evaluated and the results of the no-risk evaluation are shown in Table 2.

Table 2

Cash Flows for Example Capital Project

Time (years)	Cash Flow ($1000)	Cumulative Cash Flow ($1000)	Discounted Cash Flow ($1000)	Cumulative Discounted Cash Flow ($1000)
0	-700	-700	-700	-700
1	-900	-1600	-783	-1483
2	400	-1200	302	-1181
3	600	-600	395	-786
4	600	0	343	-443
5	900	900	447	4
6	900	1800	389	393*

Corresponds to Internal Rate of Return = 22.9%

*Net Present Worth = $393,000 at 15%, no risk

The merit of the project was evaluated on the basis of the net present worth and the internal or discounted cash flow rate of return using discrete interest compounding, a hurdle interest rate of 15% and with the risk-exponent taken as unity (i.e. no-risk). In accordance with these results the project would be expected to return a net present worth of $393000 and an internal or discounted cash flow rate of return of 22.9%. These results were then compared with those obtained when uncertainty was taken into account by means of a risk-exponent with a value other than unity. The results are presented in Table 3 at the same hurdle rate of 15% but with the risk-exponent taken as 1.2 for the expected revenues. The net present worth (cumulative discounted cash flow) under these conditions has now been reduced to $64000 and the discounted cash flow rate of return is 16% as compared to $393000 and 22.9%, respectively.

Table 3

Cash Flow and Discounted Cash Flows
Using Risk Exponent = 1.2
Hurdle Rate = 15%

Time (years)	Cash Flow ($1000)	Risk Discount Factor	Discounted Cash Flow ($1000)	Cumulative Discounted Cash Flow ($1000)
0	-700	1.0000	-700	-700
1	-900	.8696	-783	-1483
2	400	.7254	290	-1193
3	600	.5931	356	-837
4	600	.4782	287	-550
5	900	.3813	343	-207
6	900	.3012	271	64*

Corresponds to Internal Rate of Return = 16 %

*Net Present Worth = $64,000.-

It is apparent that this capital project is much less desirable when uncertainty considerations are invoked. If the projected cash flows shown in Table 2 were considered to be for two different project proposals, one with a risk-exponent of unity and the second with a risk-exponent greater than unity, it is apparent that the second project would receive a lower ranking than the first one even though all the cash flows and their timings are identical.

PROJECT RANKING EXAMPLE

A further example is presented to demonstrate how the risk-exponent can be of assistance in the ranking of competing project proposals. The cash flow profiles for three projects are presented in Table 4 and several measures of merit are shown in Table 5 for the case of a risk exponent equal to unity. Each of the three projects has a net present worth of $200000 at a hurdle rate of 20% and, in addition, the internal rates of return are equal to 25% for each of the three projects. The only major difference among the projects is the maximum amount of capital at risk which is $880000 for A as compared to $1,795,000 and $1,349,000 for projects B and C, respectively. It would appear, therefore, that project A would be the preferred choice with the ranking being A, C, B.

The picture changes, however, when uncertainty is invoked by the use of the risk-exponent or risk-time. The results of the calculations for net present worth and the internal rate of return are presented in Table 6 for risk-exponents of 1.1 and 1.2. For the case of the higher-valued risk-exponent the net present worths were negative for each of the three projects and, in addition, the internal rates of return were all less than 20%. Project C would suffer even more if a risk exponent smaller than unity would be

Table 4

Cash Flow Profiles for Project Ranking Example

Time (years)	Cash Flow (in $1000)		
	A	B	C
0	-880	-1795	-376
1	0	+667	-376
2	0	+667	-376
3	0	+667	-376
4	0	+667	-376
5	+2687	+667	+3855

Table 5

Measures of Merit for Project Ranking Example

Risk Exponent = 1.0

Profit Measures	Project		
	A	B	C
NPW at 20% hurdle rate ($1000)	200	200	200
Internal Rate of Return	25 %	25 %	25 %
Maximum Capital at Risk, Discounted at 20% ($1000)	880	1795	1349

used for the late-occuring expenditures with the value for r greater than unity being reserved only for the revenues.

Table 6

Measures of Merit for Project Ranking Example with Risk Exponent

Risk Exponent	Measures	PROJECT		
		A	B	C
1.2	Net Present Worth ($1000, 20%)	-116	-19	-175
	Internal Rate of Return, %	17.6	19.6	16.5
1.1	Net Present Worth ($1000, 20%)	41	92	11
	Internal Rate of Return, %	20.9	22.1	20.2

If a risk-exponent of 1.1 were considered to be the appropriate measure of risk for all three projects the net present worths would be positive for each of the three projects but the ranking would now be B, A, C. This is not too surprising because revenue flows into project B from the beginning of the second year whereas revenues from projects A and C do not enter until the final year. These late-arriving positive cash flows suffer a large discounting penalty due to uncertainty which is much greater than that due only to the time value of money.

CHOICE OF RISK EXPONENT VALUE

The risk-exponent approach as presented above does not give any clues as to how the appropriate value of the risk-exponents is to be selected. Because the risk-exponent is an empirical concept no hard and fast rules can be presented as to the value to be used and which would be suitable for typical situations. The ultimate selection would have to be left to the decision-maker but some guidance can be given.

Guidance in the selection can be obtained by comparing risk-exponent results with results obtained by using one of the conventional probabilistic approaches [19]. One probabilistic procedure involves the estimation of the standard deviation of the cash flow from the mean where the mean is taken to be the projected cash flow. The normal distribution function is then used to calculate the probability that the actual net present worth will reach or exceed (or not attain) some given value. Obviously there is a 50% probability that the actual net present worth will reach or exceed the mean value (or, alternatively, a 50% probability that the mean value will not be attained).

Probability distributions can be calculated for any given value of the standard deviations and such calculations were made for the capital project of Table 1. The results are presented in Table 7 for a hurdle rate of 20%. Deviations of the cash flow from the mean of 10, 15, 20 and 30% were studied. The expected or mean value, i.e. the value which would be expected to be attained with a 50% probability is, of course, $200000. One can also read from this table, for example, that for the case where a 20% standard deviation of the projected cash flows is assumed, there is a 30% probability that the actual net present worth attained will be -$9000 or less (or, alternatively, there is a 70% probability that the net present worth actually attained will reach or exceed -$9000), a 20% probability that the actual net present worth will not exceed -$136000 and only a 10% probability that the net present worth will not exceed -$312000.

Table 7

Net Present Worth Probability Distributions for Example Capital Project

Hurdle Rate = 20%

Standard Deviation of Cash Flow from the Mean, %	Net Present Worth ($1000)			
	Probability that Actual NPW will be Smaller than Indicated Value, %			
	10	20	30	50
10	-56	32	95	200
15	-184	-52	43	200
20	-312	-136	-9	200
30	-567	-304	-114	200

When the risk-exponent approach is used the uncertainties in the projected cash flows become an inherent part of the risk-exponent and the result is reflected in the calculated net present worth. Thus, one could state that the risk-exponent net present worth is the expected or mean value when the uncertainties are incorporated into the calculations by means of the risk-exponent. A direct functional relationship between the risk-exponent and the standard deviation has not yet been achieved because of the nature of the normal probability and of the discount factor expressions. The calculated values obtained for the net present worth can, however, be related to the probability distribution analyses in a semi-quantitative way.

Calculations have been made for a number of reasonable cash flow patterns and risk-exponent NPW results were compared to values obtained by the use of normal probability distributions for the cash flows. These results have shown the use of a risk-exponent equal to 1.1 are generally equivalent to the probability distribution results at the 20-30% probability level for the case of a 10% standard deviation of the cash flows from the mean. For the case of an assumed value of some 20% for the standard deviation of cash flows a value of 1.2 results in a value approximately equal to the same 20-30% probability level and a 30% deviation of the cash flows corresponds approximately to a risk-exponent value of 1.3.

Thus, a decision-maker who is willing to accept a small risk in the range of 20-30% that the calculated NPW would not be achieved would need only to specify the value of the risk-exponent in accordance with the standard deviations he would normally expect for the projected cash flow values. If he expects that the projected cash flows would have a standard deviation from the mean of some 20% he need only specify a risk-exponent equal to 1.2. The resulting risk-exponent net present worth would then be the NPW which, with a probability of 70-80%, would be attained or exceeded.

LONG-DELAYED CASH FLOWS

The decision-maker is confronted with a problem in the evaluation of capital investment proposals when cash flows can be expected over a period well beyond the usual planning or time horizon. The decision-maker can deal with this problem, for example, by cutting off from consideration all late-occurring cash flows. This would mean that those long-delayed cash flows, even though subject to a high degree of uncertainty, would be completely excluded from the net present worth calculation. Alternatively, the decision-maker could assume an unrealistically high hurdle rate so that the late-occurring cash flows, although having some effect on the NPW, would have little effect on the decision. But this alternative penalizes the earlier-occurring cash flows for which the uncertainty is lower. By using the risk-exponent the decision-maker could use a realistic hurdle rate, thereby not unfairly penalizing early-occurring cash flows, and still could include the contribution of the late-occurring cash flows in the proposal evaluation, albeit at a highly discounted value.

As an example of this effect one can note that, at a hurdle rate of 20%, a cash flow that occurred at the beginning, say, of the seventh year, would be discounted to a present worth by a discount factor equal to 0.27908 at a risk-exponent value of unity which corresponds to the usual compound interest procedure. The discount factor is almost halved to 0.15206 at a risk-exponent of 1.2 and has a value of 0.06207 at a risk-exponent of 1.4. The effect of the risk-exponent is, of course, attenuated as the value for the hurdle rate is decreased. Tables such as Table 1 can provide guidance to the decision-maker in the choice of risk-exponent.

CONCLUSIONS

A simple and straight-forward method for incorporating uncertainty into investment evaluations which could prove to be a useful tool for the decision-maker is presented in this paper. The technique addresses the uncertainty patterns of cash flows as they occur in investment or capital budgeting proposals.

Although the empiricism may be disguised or transformed, it should be recognized that attempts to deal with the problem of future and uncertain events by techniques involving modelling and probability functions are basically empirical in nature. The "risk-time" empiricism proposed here is no more empirical and does not require any more guess-work and management intuition than do other procedures for the evaluation of investment proposals under uncertainty. The proposed procedure possesses the advantages of being simple, easily understood and easily implemented by the decision-maker.

Judgment, intuition and experience will be necessary for the choice of suitable values for the risk-exponent in accordance with the nature of the proposals being considered. Further work could lead to an economic interpretation of the risk-exponent approach and thereby provide an economic foundation and rationale for the selection of the appropriate value or values for the risk-exponent for specific types of investment proposals.

REFERENCES

1- Eudemus, Baba Bathra 12[a], "Babylonian Talmud"

2- Johanan, Baba Bathra 12[b], "Babylonian Talmud"

3- English, J. M., 1984, "Project Evaluation: A Unified Approach for the Analysis of Capital Investments", Macmillan Publishing Company, New York. 269-271.

4- Fama, E. F., 1977, Risk-Adjusted Discount Rates and Capital Budgeting Under Uncertainty. J. Financial Economics, 5:3-24.

5- Treynor, J.L. and Black, F., 1976, Corporate Investment Decisions, Chapter 16 in Modern Developments in Financial Management, 310-327, Dryden Press, Hinsdale, Illinois

6- Bhattacharya, S., 1978. Project Valuation With Mean-Reverting Cash Flow Streams. Journal of Finance, 33: 1317-1331.

7- Tanchoco, J.M.A., Buck, J.R. and Leung, L.C., 1981. Modeling and Discounting of Continuous Cash Flows Under Risk. Engineering Costs and Production Economics, 5:205-216.

8- Young, D. and Contreras, L.E., 1975. Expected Present Worths of Cash Flows Under Uncertain Timing. The Engineering Economist, 20: (4), 257-268.
9- Hillier, F., 1963. The Derivation of Probabilistic Information for the Evaluation of Risky Investments, Management Science, 9:443-457.
10- Rosenthal, R.R., 1978. The Variance of Present Worth of Cash Flows Under Uncertain Timing. The Engineering Economist, 23:(3), 163-170.
11- Steece, B., 1984. Ex-Ante Measures for Evaluating Forecasting Models, Engineering Costs and Production Economics. 10:25-34.
12- Kennedy, W.J.,Jr. and Davis, R.P.,1988. Including Variability of Discrete Estimates in Engineering Economic Comparisons. Engineering Costs and Production Economics, 14:211-216.
13- Leung, L.C., Hui, Y.V. and Fleischer, G.A., 1989. On the Present-Worth Moments of Serially Correlated Cash Flows, Engineering Costs and Production Economics, 16:281-289.
14- Dean, J., 1951, "Managerial Economics", p. 568, Prentice-Hall, Englewood Cliffs, New Jersey.
15- Hicks, C.L. and Bridgwater, A.V., 1986. An Overview of Modern Project Investment Appraisal, Part 2., Process Economics International, VI, (3&4), 11-25.
16- Sinclair, C.G., 1966, Measurement of Investment Worth, Chemical and Process Engineering, March, 137-139.
17- Resnick, W. ,1981. Process Analysis and Design for Chemical Engineers, McGraw-Hill Publishing Company, New York. 229-233.

V. Quality Control Techniques

ECONOMIC DESIGN OF FRACTION DEFECTIVE CONTROL CHARTS: SIMPLIFICATION OF THE MULTIPLE ASSIGNABLE CAUSES SITUATION

by
Michael H. Peters
Quantitative Business Analysis Department
Louisiana State University
Baton Rouge, LA 70803
(504) 388-2126

ABSTRACT

Several models for the economic design of p- and np-charts have been developed where the process can be disturbed by more than one assignable cause. It has been speculated that this multiple causes situation can be converted to the simpler single cause version that yields a matching solution. In this paper, heuristics are developed for determining the matching single cause problem for two multiple causes models. These new heuristics are compared with previously suggested heuristics across a wide range of values for a variety of model parameters.

INTRODUCTION

p- and np-control charts are used extensively in production systems for the surveillance of attribute quality characteristics. Attention has been given to the economic design of p- and np-charts in recent years. The economic design of these charts involves the determination of the operating parameters that minimize a relevant cost function. Sample size, the spread of the control limits and the time between taking samples (intersample interval) are the operating parameters determined by the model used.

The development of models for this type of problem has followed two avenues. Ladany [9], Chiu [1] and Gibra [5] have formulated models where there is a single assignable cause to disturb the process. Their work is based on Duncan's article [3] on the economic design of $\bar{x}$-charts. Ladany's cost function contains sampling costs, undetected defective cost, search cost in the case of a false alarm, and search, detection and elimination cost in the case of a true alarm.

The derivation of Chiu's model was similar to the derivation of Gibra's model. Economic considerations in the Chiu formulation included profit per hour when the process is in the in-control state, profit per hour when the process is in the out-of-control state, sampling costs, false alarm cost, and true alarm cost. The objective is to minimize a loss-cost per hour function. The cost components in the Gibra cost function are sampling, undetected defectives, false alarm, and true alarm. The objective is to minimize the cost per hour. Both Gibra [6] and Chiu [2] expanded their models to the more realistic multiple assignable causes situation, i.e., more than one cause could shift the process into an out-of-control state.

Using a completely different line of development, Montgomery, et al. [10] presented a model for the multiple assignable causes situation based on Knappenberger and Grandage's [8] work on the economic design of $\bar{x}$-charts. In their approach, a Markov chain model structure was employed.

In each of the papers dealing with multiple assignable causes, it was hypothesized that a single causes version of the problem could be stated that would result in the same solution produced using the multiple causes version. Thus, the multiple causes situation could be solved using the simpler single cause version of the problem. From the designer's standpoint, fewer model parameters would have to be identified and the problem could be solved more easily.

For each of the models, it was suggested that a single out-of-control proportion defective, p_1, replace the multiple out-of-control proportions defective to create the matching single assignable cause problem. In the case of the Chiu and Gibra models, it was suggested that p_1 could be determined from the multiple causes situation using

$$p_1 = \frac{\sum_{i=1}^{s} \lambda_i d_i}{\lambda} \qquad (1)$$

where d_i = proportion defective produced when the process is disturbed by assignable cause i,

λ_i = reciprocal of the mean of the exponentially distributed occurrence times of assignable cause i,

$$\lambda = \sum_{i=1}^{s} \lambda_i ,$$

s = number of assignable causes or out-of-control states.

For the Montgomery formulation, a similar relationship was suggested for converting a multiple causes situation to a single cause problem.

$$p_1 = \sum_{i=1}^{s} P_{0i} d_i \Big/ \sum_{i=1}^{s} P_{0i} \qquad (2)$$

where P_{0i} = probability of the process shifting from the in-control state directly to out-of-control state i during the intersample interval.

P_{0i} is a function of the intersample interval, s, λ, R (production rate), and π (a parameter of the distribution of P_{0i}). As π increases, shifts to out-of-control states become more likely.

Chiu and Montgomery each illustrated their "matching" schemes for one problem. Using equation (1) to compute p_1, Gibra presented "matching" single cause solutions for ten example problems for two versions of his multiple causes model. The ranges of model parameters used in these examples were very limited. With these sparce results, it is not clear that the form of heuristic suggested by these authors is the most effective means of "matching" the single cause problem with the multiple causes problem.

The tactic similar to that manifested by equations (1) and (2) will be used to develop a matching single cause problem in this paper, i.e., find the appropriate value for the single out-of-control proportion defective. The purpose is to develop an improved means of determining the matching single out-of-control proportion defective than those proposed above. For this study, matching is defined as having a single cause version of the problem which yields the same or nearly the same minimum value of the objective function as the multiple causes version of the same problem.

METHODOLOGY

To develop the relationship to calculate the out-of-control proportion defective for the matching single cause version of a problem, a surface was fitted to the data collected in the manner described below. Because they are quite different from one another, the Chiu and Montgomery models were both selected for investigation. The Montgomery model was modified slightly for an np-chart. Although the same approach was used to generate observations, data were collected and analyzed separately for each model.

The following notation was required in the two models:

a = fixed cost of sampling (\$),
b = variable cost of sampling (\$/unit),
c = expected cost of investigating and correcting a process which is apparently out of control plus the cost of lost production (\$),
f = penalty cost of producing a defective unit (\$/unit),
t_0 = expected time to search for an assignable cause when the process is in the in-control, i.e., false alarm (hours),
t_i = expected time to search for an assignable cause i ($i = 1,2,\ldots,s$), i.e., true alarm (hours),
S_0 = expected search cost under a false alarm (\$),
S_i = expected cost to detect and eliminate assignable cause i (\$),
V_0 = profit per hour generated when the process is in-control (\$/hour),
V_i = profit per hour generated when the process has been disturbed by assignable cause i (\$/hour).

To develop relationships that would be somewhat robust,

seven of the Chiu model parameters were varied during the data collection phase while eight were varied for the Montgomery model. Their values are presented in Table for the Chiu model and Table 2 for the Montgomery model. Values used for other model parameters in the Chiu model are given at the bottom of Table 1. The numbers of

Table 1
Parameter Values for the Chiu Model

Data Group	Variables						
	a	b	$\bar{S}$	V_s	λ	d_0	Δd
I	1 10	.1 1	10 1000	0 -250	.001 .1	.01 .04/ .085*	.01 .04
II	2.5 8.5	.25 .85	175 835	-42 -208	.0175 .0835	.015/ .25* .035/ .07*	.015 .035
III	4 7	.4 .7	340 670	- 83 -167	.0340 .0607	.02/ .04* .03	.02 .03
Center Point	5.5	.55	505	-125	.0505	.025	.025

$$\bar{S} = \frac{\sum_{i=1}^{s} \lambda_i S_i}{\lambda} = 2S_0 \qquad V_i = V_0 + \frac{(V_s - V_0)}{(d_s - d_0)} \Delta d$$

$$\bar{t} = \frac{\sum_{i=1}^{s} \lambda_i t_i}{\lambda} = S\,/100 \qquad V_0 = 100 \qquad t_0 = S_0/100$$

* For this set of problem, the value of this variable was split between these values.

assignable causes for the multiple causes version of the problem were set to 4, 5, and 6, and the spacing or interval, Δd, between successive proportions defective states (d_i - d_i -1) was equal in all cases.

Table 2

Parameter Values for the Montgomery Model

Data Group	Variables							
	a	b	c	f	R	π	d_0	Δd
I	1	.1	10	1	500	.2	.01	.01
	10	1	1000	50	2000	.8	.04/ .085*	.04
II	2.5	.25	175	9.5	750	.3	.015/ .25*	.015
	8.5	.85	835	41.5	1750	.7	.035/ .07*	.035
III	4	.4	340	17.5	1000	.4	.02/ .04*	.02
	7	.7	670	33.5	1500	.6	.03	.03
Center Point	5.5	.55	505	25.5	1250	.5	.025	.025

*For this set of problems, the value of this variable was split between these values.

Rather than randomly selecting combinations of model parameters to create problems, an attempt was made to systematically explore the domain delineated by the extreme values of the model parameters. For each group of model parameters shown in Table 1 and 2, values were selected for orthogonal contrasts in a 1/16 replication of a 2^7 factorial design. For the Group I values, the design was rotated once, and the Group II and III values were rotated five times. With the center point, this means 113 experimental problems were attempted for each of the three values of s.

The general procedure for obtaining one observation in a data set was as follows:

1. Find the minimum value of the objective function for the multiple assignable causes version of the problem. In this research, the pattern search procedure outlined by Hooke and Jeeves [7] was used to obtain a local minimum. Previous work using these models has indicated that the local minimum is somewhat sensitive to the starting point given to the search procedure. Thus, the problem was solved using four different sets of starting values for sample size, acceptance number and intersample interval. These starting values were generated by taking all combinations of two sampling policies (sample size and acceptance number) with two intersample intervals. One sampling policy resulted in a low probability of indicating the process was out of control (large sample size, small acceptance number) while a second sampling policy had a relatively high probability of the same indication (small sample size, large acceptance number). One intersample interval was much larger than the other. Thus, each search was initiated from one of four diverse "corners" of the cost surface. The minimum cost from the four was selected as the solution to the problem. If this solution yielded a sample size of zero or the intersample interval became very large, i.e., do not sample, the problem was dropped from the data set.
2. In the case of the Chiu model, one model parameter for the single cause version of the problem is defined by the parameters in the multiple causes version. The conversion procedure given by Duncan [4] was used to obtain that parameter, i.e.,

$$X_1 = \sum_{i=1}^{s} \lambda_i V_i / \lambda$$

where X_1 = profit per hour generated by the process when it is in the out-of-control state of the single cause version.

3. A search was carried out to find the out-of-control proportion defective in the single cause version of the problem which gave a minimum that matched the minimum found for the multiple causes version of the problem, i.e., the objective of this search was to minimize the absolute difference between the total costs found in the two versions of the same problem. As in Step 1, four different starting points were supplied to the single cause version. If the sample size was greater than zero and the intersample interval was not approaching a very large number, the data (model parameters and the solution) were recorded as an observation.

Both sets of data were analyzed by fitting various surfaces described by first and second order combinations of process and cost parameters as the independent variables and the single cause proportion defective as the dependent variable. The general linear model procedure of the Statistical Analysis System (SAS) was used to perform this part of the analysis. The corrected R^2 for a particular fit was the primary criterion used to select the surface.

RESULTS

Based on the corrected R^2 resulting from various combinations of model parameters, the surface selected for the Chiu model is described by

$$p_1 = -.0545 + .0132b - .0000232\bar{S} + .01159s + .1021\lambda + 1.1028d_0 + 2.7418(\Delta d) + .0000948(\Delta V) \quad (3)$$

where

$\bar{S}$ = average of the expected costs of detecting and eliminating the assignable causes ($),

$\Delta V = V_i - V_{i-1}$.

The corrected R^2 was .9777. Similar to the Chiu heuristic (equation (1)), parameters that determine the location and spread of the distribution of the proportions defective produced in the multiple causes model, namely s, d_0 and Δd, are directly related to the value of p_1. An increase in either s or Δd would increase the spread of this distribution to include higher proportions defective while an increase in d_0 shifts the entire distribution to a higher range of values. This, the single cause model's out-of-control proportion defective must be increased to produce the matching solution.

What is not considered by the Chiu heuristic is the affect of other process and cost parameters on p_1. An example is λ. As λ increases, the time between shifts to producing out-of-control proportions defective decreases. This means that the time that the process spends in the higher proportion defective states increases with λ. The matching single cause model requires a larger p_1 to yield the "same" answer as the multiple causes model.

Equation (3) shows that the variable cost of sampling, b, and ΔV are directly related to p_1 while $\bar{S}$ is inversely related. Typically ΔV is a negative value, i.e., profit per hour decreases as the proportion defective increases. Therefore, the last term is negative, and p_1 decreases as ΔV increases. The reasoning to support the contributions of these variables to p_1 is not straightforward. For example, results from sensitivity analyses on Chiu's multiple causes model indicates that as the variable cost of sampling increases, the probability of stopping the process (sample results indicate that the process is out of control) and the frequency of taking samples decreases. The data from this study show that the p_1 for the single cause model must be increased to create a comparable result.

For the Montgomery model, the equation selected was

$$p_1 = -.01950 - .002485a + .007952s + 1.1740d_0 + 4.7646\,\pi(\Delta d) + 24.8618(\Delta d)^2 \quad (4)$$

The corrected R_2 was .7917. The equation for this surface is a bit more complex than the equation associated with the Chiu model since several nonlinear terms are involved. As with the Chiu model equation, major contributors to the value p_1 are factors that affect the spread and location of the distribution of proportions defective for the multiple causes model, i.e., s, Δd and d_0.

Other process parameters that affect p_1 include π. P_{0i}, used in the Montgomery heuristic to compute p_1, is a function of π. As previously stated as π increases, the probability of shifting to higher out-of-control states increases. According to equation (4) as π increases, p_1 must be increased. Thus, the impact of the process being in states producing higher proportions defective is reflected in the higher matching value for p_1.

When the fixed sampling cost, a, is increased in Montgomery's multiple causes model, sensitivity analyses reveal that the probability of stopping the process increases, but the sampling frequency decreases. The matching value of p_1 must be decreased to obtain similar behavior.

A relevant question is whether computing the matching out-of-control proportion defective for the single cause version of the problem using equation (3) or (4) is superior to the alternative proposed earlier. To compare the two approaches, the following procedure was carried out for both the Chiu and Montgomery models. Using the single cause version of the model, each problem in the respective sets of observations was solved first using the heuristic suggested by the model's originator (Case 1) to calculate the matching single out-of-control proportion defective. The same problem was then solved with the matching proportion defective being determined using either equation (3) or (4) (Case 2).

The absolute values of the difference between the minimums found in the multiple causes version of the problem and the single cause version were compared on a problem by problem basis. The results of these comparisons are summarized in Table 3. The absolute value of the deviation from the multiple causes minimum for Case 1 is denoted by Dev_1; similarly the absolute value of the deviation for Case 2 is denoted by Dev_2.

For the Chiu model, 71.43% of the solutions to the single cause problem developed using equation (3) gave a smaller deviation from the multiple causes cost solution than the solution to the single cause problem generated using equation (1). The Case 2 solution was better than (smaller deviation) the Case 1 solution 86.83% of the time with the Montgomery model. Column 4 contains the average difference of the deviation from the multiple causes minimum with the single cause version determined under Case 1 and the similarly defined deviation when Case 2 was used to determined the single cause version. For both models this average is positive and significantly different from zero ($\alpha < .0001$)indicating that the fitted equations are an improvement over the heuristics suggested by the models' originators. The comparison of the two approaches was related to the multiple causes version of the problem by finding the percentage that the absolute value

Table 3
Results of Case 1 vs. Case 2 Comparisons

Model	(1) Number of Assignable Causes	(2) N	(3) Number of Observations Where $Dev_2 < Dev_1$ [%]	Averages		
				(4) Dev_1-Dev_2 ($)	(5) $\frac{Dev_2}{\text{Total Cost}}$ X100 (%)	(6) $\frac{(Dev_1-Dev_2)}{\text{Total Cost}}$ X100 (%)
Chiu	4	105	73 [69.52]	.1531	2.3032	-.1257*
	5	98	73 [74.49]	.2547	1.4067	.8470
	6	91	64 [70.33]	.2030	1.8721	.6482
	4, 5 and 6	294	210 [71.43]	.2024	1.8709	.4380
Montgomery	4	76	61 [80.26]	.0159	.9970	1.3767
	5	81	73 [90.12]	.0233	1.3754	.8304
	6	86	77 [89.53]	.0199	1.2026	.7918
	4, 5 and 6	243	211 [86.83]	.0198	.9876	1.1959

*Accept H_0 (α >.05)

of the deviation was of the multiple causes minimum cost solution for Case 2, and the percentage that the difference in the absolute deviations (Dev_1-Dev_2) was of the same minimum cost. The averages of these percentages are given in columns 5 and 6, respectively, of Table 3. Except for the Chiu model problems where $s=4$, these averages were statistically different from zero ($\alpha<.05$).

Data from observations obtained using each model were ranked by this last measure in descending order to see if there was any pattern among the model parameters for the observations where Case 2 did not yield a smaller deviation than Case 1. In the case of the Chiu model, it appeared that the problems with the largest percentage difference in deviations had most, if not all, of the model parameters at one of their extreme values, e.g., this was the case in 15 out of the first 20 problems. For this subset of 15 cases, the only pattern for the model parameters was that the interval between the proportions defective was at its minimum value in all but one case. A similar analysis of the data from the Montgomery model shows that in 40.62% of the cases where the percentage difference in deviations was negative, the model parameters were at one of their extreme values. Only 13.60% of the problems in the problem set had all of the model parameters at one of their extreme values. Examination of the subset of problems where the parameters were at an extreme value shows that when b and c are at their minimum values, and when f (cost of an undetected defect) and π are at their maximum values, the largest negative percentage difference occurs.

The analysis used to generate the data in Table 3 was performed with problems containing all model parameters at extreme values removed. The results are given in Table 4. When compared with the data appearing in Table 3, there is a significant improvement in the averages (columns 4, 5 and 6) with the Chiu model, but the percentage of cases where $Dev_2 < Dev_1$ did not change a great deal. For the Montgomery model, the opposite situation occurred.

CONCLUSION

The results indicate that equations (3) and (4) offer an improved means of calculating the matching out-of-control proportion defective over equations (1) and (2), respectively.

Table 4
Results of Case 1 vs. Case 2 Comparisons
With Extreme Point Problems Removed

Model	(1) Number of Assignable Causes	(2) N	(3) Number of Observations Where $Dev_2 < Dev_1$ [%]	Averages		
				(4) $Dev_1 - Dev_2$ ($)	(5) $\frac{Dev_2}{\text{Total Cost}} X100$ (%)	(6) $\frac{(Dev_1 - Dev_2)}{\text{Total Cost}} X100$ (%)
Chiu	4	91	68 [74.72]	.2819	1.3648	.7659
	5	85	65 [76.47]	.3382	.9476	1.1612
	6	78	59 [75.64]	.3261	1.3532	1.0461
	4, 5 and 6	254	192 [75.59]	.3143	1.2216	.9843
Montgomery	4	69	58 [85.51]	.0190	1.2907	1.1877
	5	74	70 [94.59]	.0264	.7429	1.5565
	6	78	74 [94.87]	.0227	.7025	1.3486
	4, 5 and 6	221	202 [91.40]	.0228	.8997	1.3680

For both heuristics for both models, the distribution of the proportion defective on the multiple causes version are primary contributions to the matching single out-of control value. The improved heuristics include the affects of additional process and cost parameters. Either improved equation provides the designer with a relatively simple means of converting the complex multiple causes situation into a matching single cause problem. Although only one distribution of out-of-control proportions defective was investigated, these equations should apply across a variety of distributions particularly if we accept Duncan's [4] conclusion that the cost function's minimum appears more or less invariant with respect to the shape of the prior distribution of out-of-control proportion defective for the X-chart.

Modification of the value of the out-of-control state is only one tactic for simplifying the multiple causes situation. Alternative ways of restructuring are topics for future research.

REFERENCES

[1]. Chiu, W.K., "Economic Design of Attribute Control Charts," Technometrics, 17, 81-87, (1975).

[2]. ______, "Economic Design of np Charts for Processes Subject to a Multiplicity of Assignable Causes," Management Science, 23, 404-411, (1976).

[3]. Duncan, A.J., "The Economic Design for X-charts Used to Maintain Current Control of a Process," J. Amer. Statist. Assn., 5, 228-242, (1956).

[4]. ______, "The Economic Design of X-charts When There is a Multiplicity of Assignable Causes," J.Amer. Statist. Assn., 66, 107-121, (1971).

[5]. Gibra, I.N., "Economically Optimal Determination of the Parameters of np-Control Chart," Journal of Quality Technology, 10, 12-19, (1978).

[6]. Gibra, I.N., "Economic Design of Attribute Control Charts for Multiple Assignable Causes," Journal of Quality Technology, 13, 93-99 (1981).

[7]. Hooke, R. and Jeeves, T.A., "Direct Search Solution of Numerical and Statistical Problems," Journal of the Association for Computing Machinery, 8, 212-229 (1961).

[8]. Knappenberger, H.A. and Grandage, A.H., "Minimum

Cost Quality Control Tests," AIIE Transactions, 1, 24-32, (1969).
[9]. Ladany, S.P., "Optimal Use of Control Charts for Controlling Current Production," Management Science, 19, 763-772, (1973).
[10]. Montgomery, D.C., Heikes, R.G. and Mance, J.F., "Economic Design of Fraction Defective Control Charts," Management Science, 21, 1272-1284, (1975).

VI. Warranties and Repair Parts Costing

An Active Decision Support System for Warranty Cost Estimation

Bin-Shan Lin

ABSTRACT

This study develops an active decision analysis model-base which can be implemented as an ADSS to gain insight about warranty cost estimation. Under different warranty policies, a framework which leads to expressing expected warranty costs in terms of the renewal function is discussed. We also consider how to classify and integrate design considerations into the renewal framework for warranty cost estimation.

1. INTRODUCTION

A warranty can be defined as a contractual obligation under which a manufacturer must agree to repair or replace a commercial or a consumer product should it fail before a certain length of time (warranty period). The elapsed time may be measured in calendar time units (months or days) or in usage units (miles, minutes of operations, etc.). A warranty is associated with almost all products, whether sold directly to the customer or to a producer for assembly into a consumer product. The nature and extent of the warranty affect the sales, market share, costs and profits of business. As such, it is an integral part of satisfactions which the consumer receives when (s)he purchases a product.

Warranties have a significant effect on initial equipment operating cost. This effect should be considered and estimated when selecting a new product. Often, manufacturers use warranties as a means of advertising the quality of products to accomplish a variety of goals such as quality assurance and product promotion. Warranties also represent powerful and pervasive "risk-influencing mechanisms" (Shimp and Bearden,

1982). In situations involving large and costly pieces of equipment, the effect of warranty cost can become very critical.

Warranty cost estimation is of importance to manufacturers. The knowledge of total expected warranty liability will help manufacturers plan operations more effectively, since an accurate knowledge of warranty costs allows more accurate profit expectations. Failure to record estimated warranty liability will overestimate income and underestimate liability.

There have been few attempts to devise decision support mechanism and systems which can aid in the formulation and implementation of warranty policies. The purpose of this paper is twofold: first to review the work on developing decision support system (DSS) with analytical models intended to provide insights into the warranty cost estimation to bring out some of these implications. We attempt to develop a framework that is based conceptually on what has been learned about problem solving processes and to build on the emerging concept of active decision support system (ADSS) as a network of system interfaces. We also consider how to classify and integrate design considerations into a framework for warranty cost estimation.

The paper is organized as follows. Section 2 contains a review of the DSS with particular focus on ADSS. In Section 3, the analytical models for warranty cost estimation problems are addressed. Section 4 addresses the renewal framework for the warranty cost estimation. Section 5 presents an example with an illustration of the renewal model. Section 6 describes the ADSS structure for warranty cost estimation. Finally, conclusions and new research directions are discussed.

2. ACTIVE DECISION SUPPORT SYSTEM

A DSS is a computer-based support system for decision makers who deal with unstructured or semi-structured problems. A DSS consists of three components: a data base, a model base, and a user interface (Spague, 1980). A data base provides the data required for decision making, a model base supplies appropriate models, and a user interface gives a channel so that users can communicate with the system.

An ADSS requires that the computer and user work as partners in the problem-solving process (Manhein, 1988). The ADSS concept differs from conventional DSS approaches in that it is learning-based (Dolk and Kridel, 1990) rather than being centered in a system design context (Sprague and Carlson, 1982). To integrate a system into a strategy, the design issues of the system should be focused on the problem-solving processes within the organization. More importantly, we use the ADSS concept to enhance the decision-making process in developing an active warranty cost system for manufacturers.

Overall, the objective of an ADSS must be that of achieving an information-based advantage, both in support of the solving process and in the support of control issues. A good system environment should be hospitable not only to professional operators, but also to managers. Many decision situations faced by top-level managers are one-of-a-kind, one-shot-occurrence problems. The ill-structuredness of such decisions aggravates the problem of limited resources available to a business executive. A DSS emerges out of the investigation into the need for tools to solve these unstructured or semi-structured problems for which it is difficult to prescribe an algorithmic solution (Alter, 1980). Since ill-structured situations are common at the strategic level of an organization, much DSS research has been concerned with support for strategic planning and decision making at top management.

The problem of estimating a warranty cost is very complex, as it involves many decision variables, sources of uncertainty, and multiple objectives of values. For example, the design of a warranty involves establishing conditions of compensation and the period of warranty. The selection of a warranty policy is determined by competition, sales effects, and costs (Ritchken and Tapiero, 1986). The costs are determined by the rebate policy and the hazard function of the product.

A useful decision-aiding framework must be both comprehensive and friendly. An ADSS is designed for implementation in a friendly, interactive problem-solving environment. Given its capacity for substantial computation, the ADSS can propose candidate actions and optimize among them, thereby creating a recommended course of action for top-level approval. An ADSS provides managers with the capability to participate in the procedures of estimating and control task. An ADSS is required that quantifies, identifies, prioritizes, and evaluates issues and integrates them into a friendly framework.

3. ANALYTICAL MODELS

This section reviews the literature of analytical models of warranty cost estimation. Much of the previous literature has dealt with estimation of warranty cost to develop rules for better structuring the warranty problem. If the warranty period is long, present value analysis is incorporated to account for time value of money (Amato, Anderson, and Harvey, 1976). Lowerre (1968) developed a formulation of the warranty reserve fund for repairable products based on choosing a selected percentage of the revenue from each sale. Menke (1969) suggested a framework with irreparable products and the rebate strategy being a fixed period, either linear pro rata or lump sum, for any failure within the warranty period. Expected repair cost for the consumer over the life of the product is considered by Heschel (1971). Assuming an exponential failure distribution, prediction intervals for warranties were developed by Patankar and Worm (1981). A goal programming approach for the warranty cost estimation was suggested by Mitra and Patankar (1988).

Warranty cost estimation requires a number of technical and behavioral decisions. One important technical decision is the design of the warranty cost. The warranty cost design involves (1) consideration in purpose, (2) consideration in objectives, and (3) selection of policy types with determination of warranty period. The design decision is important, since it affects all levels of the firm. For example, the design parameters of the warranty cost can affect the long-term image of the firm's product, a top-level management concern. Middle-level managers may be interested in achieving a minimum cost, so designs must be cost effective. The actual users of the active DSS for warranty cost estimation are interested in designs that are simple to understand, friendly to implement and utilize.

A. Consideration in Purposes

Warranties are offered by manufacturers either for their own promotional or for protective purposes (Udell and Anderson, 1968). The first design consideration is to determine whether to adopt a promotional or protective approach to warranties. The promotional warranty is viewed as policy instruments to improve market performance or to alter consumer perceptions on quality (Spence, 1977). The ultimate decision on whether or not to employ a promotional warranty depends on the estimated ability of the warranty to increase sales and profits. If the expected warranty costs are more than covered by the gross contribution profit, the use of a promotional warranty will be profitable. In such situations, estimating warranty costs is very important since it could drastically affect profitability.

The protective warranty is employed to guard the manufacturer from unreasonable claims of consumers. Such warranties limit the manufacturer's responsibility to defects of products during a limited period of time. In this situation, estimating the warranty cost is related to estimating the warranty period.

B. Consideration in Objectives

Conventional work on this topic has mostly dealt

with formulating a linear programming with a single objective: the minimization of the warranty reserve cost or the maximization of profit. A more realistic approach involves several objectives, some of which might be conflicting to others. Mitra and Patankar (1988) showed a warranty example with three objectives: minimization of warranty reserve cost, offering a minimum level of warranty time based, and capturing a minimum specified market share of the product.

C. Selection of Policy Types

Nguyen and Murthy (1984) suggested two types of warranty policies: Policy 1 and Policy 2. Under Policy 1, the manufacturer will replace the item without any charge at each failure within a certain time W, where W is the free replacement warranty duration. If failure occurs after W, but before W+T, then the customer purchases a new item with a new warranty at a fraction of the replacement price P, where T is the pro rata warranty duration. Under Policy 2, however, the warranty is renewed after each replacement. When W=0, Policy 2 is just the pro rata policy. When W=T=0, Policy 2 is the pro rata with free replacement. Let R be the replacement cost to the manufacturer and P be the replacement price that the consumer pays. Let L be the product life cycle and $L \geq W+T$. For planning and tax purposes, L is usually assumed to be a constant. Nguyen and Murthy (1984) suggested that the expected warranty costs, C_1 and C_2, to the manufacturer under Policies 1 and 2 respectively as follows:

$$C_1 = RH(L)-P\{H(L)-T^{-1}\int_0^T F(t)dt-T^{-1}\int_0^T (T-t) H(L-W-t)dF(t)\}, \quad (1)$$

$$C_2 = RH(L)-P\{H(L)-\int_0^W H(L-t)dF(t)-T^{-1}\int_0^T F(W+t)- T^{-1}\int_0^T (T-t)H(L-W-t)dF(t+W)\}, \quad (2)$$

where F(t) is the lifetime distribution, R is the replacement cost to the manufacturer, and H(t) is the renewal function. Blischke and Scheuer (1975) derived the expected costs for a free replacement warranty by using a straight line approximation to the renewal function H(t). Mamer (1982) addressed an alternative approximation to the expected warranty cost.

The subject of interest is the expected number of replacement items, founded from H(t), in order to estimate the cost of a given warranty problem.

4. A RENEWAL FRAMEWORK FOR WARRANTY COST ESTIMATION

Reliability studies often are interested in the number of failures of the unit over a time period. The renewal models have been useful for analysis of warranties when it is important to have a good measure of the expected number of failures (Frees, 1986a,b). A renewal process is a sequence $\{X_1, X_2, \ldots\}$ of mutually independently and identically distributed random variables (Feller, 1971). In warranty applications these random variables represent the successive lifetimes of a unit or system that, upon failure, is replaced by a new one or is overhauled to as-new condition.

To define the renewal function, let X_1, X_2, distributed random variables with the distribution function F(x), $x \geq 0$, which has mean μ and variance σ^2. The renewal counting process $\{N(t): t \geq 0\}$ is introduced. N(t) is the number of failures that have occurred by time t. Consider the sum $S_k = X_1 + \ldots + X_k$, $k \geq 1$ of k failure times. The renewal function H(t) $(t \geq 0)$ is defined as,

$$H(t) = \sum_{k=1}^{\infty} F^k(t), \tag{3}$$

where $F^k(t) = Pr(S_k \leq t)$ is the k-fold convolution of F for $k \geq 1$. The function H(t) is the expected numbers of renewals in an interval [0,t] for a renewal process with underlying lifetime distribution F. If

F is completely known, then H(t) can be computed in principle. However, in most cases F is unknown or the general form of F might be known but the parameters of the distribution are unknown. In those cases H(t) has to be estimated from lifetime data. The renewal function might be estimated parametrically if the general form of F is known, or nonparametrically if the form of F is unknown.

It can be shown that the renewal function H(t) satisfies the so called renewal equation

$$H(t) = F(t) + \int_0^t H(t-y)dF(y), \tag{4}$$

and the renewal density function $h(t) = H'(t)$ satisfies

$$h(t) = f(t) + \int_0^t h(t-y)f(y)dy, \tag{5}$$

where f(y) is the probability density function of the failure times. If h(t) is known, then (5) may be used to compute the renewal function H(t) by

$$H(t) = \int_0^t h(u)du. \tag{6}$$

Renewal functions are known in closed form only in a few special cases. In general, evaluating the renewal function for a given lifetime distribution involves summing an infinite series of convolution integrals (Cleroux and McConalogue, 1976; McConalogue, 1981), numerically solving the renewal integral equation (Soland, 1969; Deligonul and Bilgen 1984), or numerically inverting a Laplace transform and some form of power series expansion (Smith and Leadbetter, 1963; Lomnicki, 1966; Jaquette, 1972).

An important asymptotic result in renewal theory is suggested by the relationship (Feller, 1971)

$$\lim_{t\to\infty} H(t) - t/\mu = \sigma^2/(2\mu^2) - ½ . \quad (7)$$

This asymptotic approximation is particularly attractive where F is not completely specifiable, as the renewal function H(t) depends asymptotically only upon the first two moments. Specifically, when the time of interest is large, asymptotic expressions for the renewal function are used. The estimator based on (7) is denoted as $\hat{H}_{na}(t)$, which is suggested to be used in estimating the C_1 and C_2 under the assumption of a known distribution (Frees and Nam, 1988).

The basic renewal function and some quantities have been evaluated by a direct solution (Soland 1968, 1969). Soland's algorithm is to first compute the renewal density function h(t) in (3) by approximating the integral with a finite sum of h(y) for y<t and then compute the renewal function H(t) by numerical integration of expressions (4). Therefore, given estimates for the parameters of f(y), one can compute the renewal function H(t) numerically.

The algorithm of Cleroux and McConalogue (1976) generates highly accurate piecewise polynomial approximations to convolutions of probability distributions that are twice differentiable. The essence of this algorithm is a cubic-spline representation of the form $F^k(t)$ in (3), where F is bounded. However, the major limitation of the Cleroux-McConalogue's algorithm is that it cannot be applied to the Weibull and the gamma distributions with decreasing hazard rate (Baxter, Scheuer, Blischke, and McConalogue 1981). McConalogue (1981) generalized this algorithm, permitting its application to a subclass of those distribution functions F with a shape parameter no less than 0.5. It has been used in Baxter et al. (1981) to compute extensive renewal tables for a number of distributions for a wide range of values of the shape parameter, with a reported accuracy to four to six decimal figures.

Soland's algorithm is much simpler, and an extensive numerical study and comparison with tables by Baxter et al. showed that Soland's algorithm performs well as long as the hazard rate is not rapidly decreasing. When the hazard rate decreases too fast, Soland's algorithm does not perform well. For instance, for the gamma distribution with the shape parameter α=0.5 Soland's algorithm is not efficient (Schneider, Lin, and O'Cinneide, 1990).

In many situations an underlying parametric form of the distribution function F(x) is not known. Thus, it is desirable to have a nonparametric estimate of the renewal function, since we may not wish to assume any specific form of F(x). Frees (1986a,b) and Schneider, Lin, and O'Cinneide (1990) suggested several nonparametric estimators for the renewal function for complete data.

In many real cases, however, censoring of the lifetime might occur. The manufacturer of the product has only a knowledge of the time to failure if it occurs before expiration of the warranty, i.e., he has censored data. The random censoring model is often used in reliability analyses and survival analyses (Miller and Halpern 1982; Lawless 1982). This is examplified by some sectors of manufacturers that offer warranties on every new product. In such a situation, Y_i is the random variable of interest. The products will have run within a given warranty period. We also assumed that the manufacturer has a random sample of observations of the product's lifetime, denoted as Y_1, ..., Y_n, available for estimation. The random censorship which is a result of the random starting times can be described by the following model. The sample is randomly censored, i.e., a random sample (Y_1, ..., Y_n) is drawn from a population with cumulative distribution function (cdf) F(y), and a random sample (C_1, ..., C_n) is drawn from cdf. G(C), where C_i (i= 1, ..., n) are the random censoring times. The observed times are then

$$X_i = G_i Y_i + (1 - G_i)C_i, \tag{8}$$

where

$$G_i = \begin{cases} 1 \text{ if } Y_i \le C_i \\ 0 \text{ if } Y_i > C_i. \end{cases}$$

Therefore, we observe the pairs (X_i, G_i), where $G_i = 0$ means that Y_i is censored and $G_i = 1$ means that Y_i is a failure time.

Based on (7) if the data are not censored, the mean μ and variance σ^2 can be conveniently estimated by the sample mean and sample variance. However, if data are randomly censored, then nonparametric estimation of μ and σ^2 becomes more complicated. Schneider et al. (1990) used the Susarla-Tsai-Van Ryzin (1984) estimator for the mean $\hat{\mu}_n$ and the Schneider-Weissfeld (1986) estimator for the variance $\hat{\sigma}^2_n$.

The nonparametric estimates $\hat{\mu}_n$ and $\hat{\sigma}^2_n$ are used in the equation (5) to obtain the nonparametric asymptotic estimator $\hat{H}_{na}(t)$ of the renewal function. Then we have

$$\hat{H}_{na}(t) = t/\hat{\mu}_n + \hat{\sigma}^2_n/(2\hat{\mu}^2_n) - ½ . \tag{9}$$

Lin (1988) suggested solving equation (4) where F(t) is replaced by the Kaplan-Meier (1958) estimate $\hat{F}$. This nonparametric estimator is denoted as $\hat{H}_{nk}(t)$. The method used to compute $\hat{H}_{nk}(t)$ is similar to the approach used to compute in the nonparametric empirical estimator $\hat{H}_{ne}(t)$ (Schneider, Lin, and O'Cinneide, 1990). Therefore, we have

$$\hat{H}_{nk}(t) = \hat{F}_k(t) + \int_0^t \hat{H}_{nk}(t-x)d\hat{F}_k(x). \tag{10}$$

In summary, there are three approaches to estimating the renewal function: (1) the lifetime distribution F is known and all parameters are known, (2) the lifetime distribution F is known but the parameters are unknown, and (3) the lifetime distribution F is unknown. Under the first situation, both Soland's algorithm and Cleroux-McConalogue's algorithm can be used to solve the renewal function numerically. Under the second situation, both Soland's algorithm and Cleroux-McConalogue's algorithm can be employed to estimate the renewal function parametrically for either censored data or uncensored data. Also, the asymptotic formula of the renewal function can be used to estimate H(t) via the estimates of mean and variance. Lin (1988) suggested that if the lifetime distribution is known, then the parametric estimator $\hat{H}_{pi}$ can be used. Under the third situation, the nonparametric empirical estimator can be used for the uncensored case as long as the hazard rate is not rapidly decreasing, while the Kaplan-Meier estimator or the nonparametric asymptotic estimator can be used for the censored case. Even though neither the nonparametric empirical nor the nonparametric asymptotic estimators is not consistent, the mean squared errors are not higher than the MSEs of the parametric estimator $\hat{H}_{pi}$. Another advantage of nonparametric estimators is that parametric estimators have to involve an iterative procedure for finding the MLEs.

Under some situations, for example, a warranty cost analysis, t is small relatively to the mean failure time of the product, there are some limitations in using $\hat{H}_{na}$. When t is very small and the lifetime distribution is unknown, the Kaplan-Meier estimator, $\hat{H}_{nk}$, is employed.

As shown in the simulation results (Lin, 1988),

both the hazard rate and the censoring distribution are crucial factors in determining estimators of the renewal function. Unfortunately, both Soland's and Cleroux-McConalogue's algorithms have some limitations to their use under the decreasing hazard rate (DHR) case. When the lifetime distribution has a fast DHR, it is very difficult and sometimes impossible to solve (2) or (3). Under such circumstances, the asymptotic estimator, $\hat{H}_{na}$, is suggested if t is not too small.

5. AN EXAMPLE

The proposed model is now demonstrated in the context of an application. We consider the 107 observed lifetimes of a unit of electronic ground support equipment given by Juran and Gryna (1980). For illustration purposes we censored the data randomly, using a uniform distribution in the interval [0, 200] which resulted in 26% censored observations. A plot of the hazard rate of the original data suggests an early failure period of about 20 hours and a wear-out period of about 100 hours. The period between 20 and 100 hours is fairly constant. We estimated the expected number of failures for 20 hours and 100 hours.

Using the two-parameter Weibull distribution with density

$$f(y) = \alpha\beta^{-\alpha}y^{\alpha-1}\exp[-(y/\beta)^{\alpha}]. \tag{11}$$

we obtain the maximum likelihood estimates $\hat{\alpha}$ = 64.194 and $\hat{\beta}$ = 0.945. Based on these estimates, we used Soland's algorithm to compute an estimate of the expected number of failures for 20 hours and 100 hours. We also calculated an estimate based on the lognormal distribution with density

$$f(y) = (y\sigma\sqrt{2\pi})^{-1}\exp[-(\log y^{-\theta})^2/(2\sigma^2)]. \tag{12}$$

The maximum likelihood estimates of the

parameters are $\hat{\theta} = 3.602$ and $\hat{\sigma} = 1.516$. Soland's algorithm is used again to obtain the estimates. The results are presented in Table 1.

TABLE 1

Estimates of the Renewal Function for Electronic Ground Support Equipment

	Parametric		Nonparametric	
	$\hat{H}_{pi}$		$\hat{H}_{na}$	$\hat{H}_{nk}$
t	lognormal	Weibull		
20	0.420	0.328	0.123	0.427
100	1.470	1.385	1.485	1.388

The nonparametric estimate $\hat{H}_{na}$ of the renewal function is based on the nonparametric estimates of the mean and variance, which are $\hat{\mu} = 58.73$ and $\hat{\sigma}^2 = 194.67$, respectively. Equation (9) can now be used to compute the estimates which are depicted in Table 1.

To compute the estimator $\hat{H}_{nk}(t)$ we first multiply the data by 100 to obtain discrete values. After obtaining the empirical distribution function using the Kaplan-Meier estimator, we finally compute $\hat{H}_{nk}(t)$ recursively. All estimates are similar except $\hat{H}_{na}$ for $t = 20$. The asymptotic estimator does not seem to perform well for small t values in the warranty case.

6. THE STRUCTURE OF ADSS

The use of a network of subsystem interfaces for the design of an active system is based on the Bonczek, et al. (1981) model of the DSS structure: a language system, a problem process system, and a knowledge system. The structure of an

ADSS includes three components: process managers, history processor, and display interfaces (Dolk and Kridel, 1990). Based on these concepts, an architectural construct for the design of a warranty estimation system can be developed, as shown in Figure 1. The structure consists of four major systems, most of which find their counterparts in the Bonczek et al. and Dolk and Kridel models. The four major systems are user interface system (UIS), cost estimation system (CES), history record system (HRS), and knowledge system (KS). Figure 1 illustrates the structure.

The first component is the UIS. In order to achieve adaptive, easy-to-use interface, a modified generic model, suggested by Dos Santos and Holsapple (1989), is employed to build the UIS. A language subsystem (LS) and a presentation subsystem (PS) are included in the UIS. The LS is the means by which the decision maker communicates with the system, while the PS refers to the communications from the system to the decision maker (Dos Santos and Holsapple, 1989). User requests are stated in terms of the LS. Normally, these are quests for lifetime data retrieval or requests for a computation for cost estimation. ADSS may stress exploratory data retrieval and may be oriented toward model execution. A typical UIS component would include a goal decomposition model to represent the renewal framework in a hierarchical decision structure for the warranty cost estimation procedure.

FIGURE 1

The Structure of an ADSS for Warranty Cost Estimation

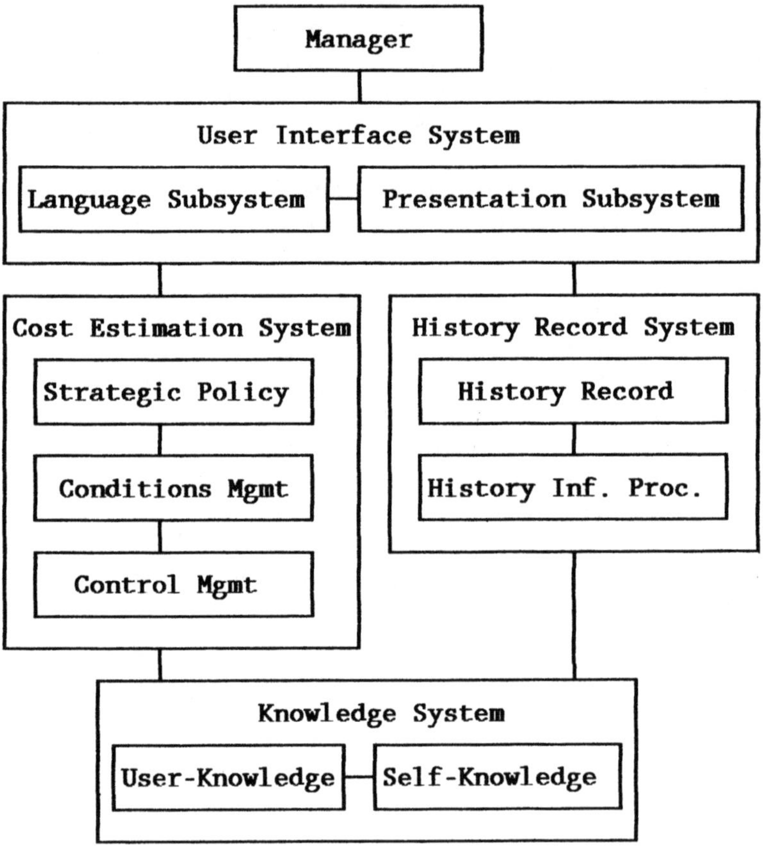

Second, a model for CES includes three subsystems: strategic policy, conditions management, and control management. The CES can provide appropriate cost planning and control models as decision aids. These models can be adapted to the situations as strategic processing in CES. The CES is a dynamic software system controlling ADSS behavior. By its cost information collection ability a CES accepts strings of symbols according to its LS syntax and acquires strings of symbols organized according to its KS representation methods.

Third, the HRS consists of two components: a history record (HR) and a history inference processor (HIP). The HR is used to journal user inputs, commands, and planning results in a history record. The HIP is employed to identify an estimation procedure or model of the user's image of the warranty cost problem from the history record.

Fourth, the KS has two parts: user-knowledge and self-knowledge. A user-knowledge based component possesses two types of knowledge: (1) domain-specific cost knowledge (K_d) and (2) support system knowledge (K_s). A decision maker uses the two types of knowledge to interact with the ADSS in the course of making decisions. In an ADSS, K_d could consist of warranty cost knowledge that is described procedure for reasoning and for presentation. K_s possesses knowledge pertaining to the use of the ADSS's capabilities. A self-knowledge based component possesses knowledge about itself, i.e., it knows what it knows about the censoring models. In an active system a self-knowledge based component is used to interpret messages/outputs and to help the decision maker obtain knowledge relevant to the cost estimation tasks.

Given a warranty cost estimation problem (stated with the LS) and a problem domain knowledge (represented in the KS), the CES attempts to derive a solution that supports. facilitates, and enhances the process of cost estimation. Several production rules could be developed within the renewal framework. The conditions include (1) the number of failed products available, (2) the product lifetime distribution, (3) censored data or uncensored data, (4) hazard rate of the product, (5) the observation time.

The ADSS requires some knowledge about the warranty cost domain. Much of the value of the ADSS results from this warranty-specific knowledge which typically includes many facts that a decision maker has neither capacity, time, opportunity, nor inclination to acquire and retain. But, certain subsets of the knowledge are essential for effective cost estimation when confronted with a particular warranty problem. It is crucial to obtain the knowledge in the KS about which methods or models to estimate and to have insight into how warranty policy changes with changes in product lifetime, different types of censored data arise in lifetime experiments, etc.

7. CONCLUSIONS

This paper attempts to develop an ADSS for aiding the formulation of warranty cost estimation. The system can usefully support warranty decisions and provide insights about the complex relationships between variables in decision categories, which in practice are often considered in isolation. The user interface system, the cost estimation system, the history record system, and the knowledge system are integrated into a knowledge-based system implemented in every estimation procedure.

Further research is suggested in a number of directions. Work on renewal models which support the central framework and link other cost estimation models is needed. Further research in applying rule-based techniques is warranted. Finally, we conclude that many applications are possible now, but that full realization of the promise of ADSS awaits a number of advances in representation and reasoning capabilities, together with integration of warranty strategy into a unified framework.

REFERENCES

[1] Alter, S.L. Decision Support Systems: Current Practice and Continuing Challenges, Massachusetts, Addison-Wesley, 1980.

[2] Amato, H.N. Anderson, E.E. and Harvey, D.W., "A General Model of Future Period Warranty Costs," The Accounting Review, 51, 4, (1976), pp.854-862.

[3] Anderson, E.E. "Product Price and Warranty Terms: An Optimization Model," Operations Research Quarterly, 28, (1977), pp.739-741.

[4] Baxter, L.A. Scheuer, E.M. Blischke, W.R. and McConalogue, D.J. "Renewal Tables: Tables of Functions Arising in Renewal Theory," Technical Report, University of Southern California, 1981.

[5] Blischke, W.R. and Scheuer, E.M. "Calculation of the Cost of Warranty Policies As a Function of Estimated Life Distribution," Naval Research Logistics Quarterly 22, (1975), pp.681-696.

[6] Bonczek, R.H. Holsapple, C.W. and Whinston, A.B. Foundations of Decision Support Systems, New York, Academic Press, 1981.

[7] Cleroux, R. and McConalogue, D.J. "An Numerical Algorithm for Recursively-Defined Convolution Integrals Involving Distribution Functions," Management Science, 22, (1976), pp.1138-1146.

[8] Deligonul, Z.S. and Bilgen, S. "Solution of the Volterra Equation of Renewal Theory With the Galerkin Technique Using Cubic Splines," Journal of Statistical Computation and Simulation, 20, (1984), pp.37-45.

[9] Dolk, D.R. and D.J. Kridel. "An Active Modeling System for Econometrics Analysis," forthcoming in Decision Support Systems, (1990).

[10] Dos Santos, B.L. and Holsapple, C.W. "A Framework for Designing Adaptive DSS Interfaces," Decision Support Systems, 5, (1989), pp.1-11.

[11] Feller, W. An Introduction to Probability Theory and Its Applications, New York, John Wiley, 1971.

[12] Frees, E.W. "Warranty Analysis and Renewal Function Estimation," Naval Research Logistics Quarterly, 33, (1986a), pp.361-372.

[13] Frees, E.W. "Nonparametric Renewal Function Estimation," Annals of Statistics, 14, 4, (1986b), pp.1366-1378.

[14] Frees, E.W. and Nam, S.H. "Approximating Expected Warranty Costs," Management Science, 34, 12, (1988), pp.1441-1449.

[15] Gunn, T.G. "The Mechanization of Design and Manufacturing," Scientific American, 247, (1982), pp.115-130.

[16] Heschel, M.S. "How Much is a Guarantee Worth?" Industrial Engineering, 3, (1971), pp.14-15.

[17] Jaquette, D.L. "Approximations to the Renewal Function m(t)," Operations Research, 20, 3, (1972), pp.722-727.

[18] Juran, J. and Gryna, F. Quality Planning and Analysis, New York, McGraw-Hill, 1980.

[19] Kaplan, E.L. and Meier, P. "Nonparametric Estimation from Incomplete Observation," Journal of the American Statistical Association, 53, (1958), pp.457-481.

[20] King, W.R. "Strategic Management Decision Support Systems," Business Policy International Symposium, University of Texas at Arlington, 1983.

[21] Lawless, J.F. Statistical Models and Methods for Lifetime Data, New York, John Willey & Sons, 1982.

[22] Lin, B.S. "Estimation of the Renewal Function," Ph.D. Dissertation, Louisiana State University, Baton Rouge, 1988.

[23] Lomnicki, Z.A. "A Note on the Weibull Renewal Process," Biometrika, 53, (1966), pp.375-381.

[24] Lowerre, J.M. "On Warranties," Journal of Industrial Engineering, 19, 3, (1968), pp.359-360.

[25] Mamer, J.W. "Cost Analysis of Pro Rata and Free-Replacement Warranties," Naval Research Logistics Quarterly, 29, (1982), pp.345-356.

[26] Manheim,M. "An Architecture for Active DSS," Proceedings of the 21st HICSS, IEEE Computer Society, 3, (1988), pp. 356-365.

[27] McConalogue,D.J. "Numerical Treatment of Convolution Integrals With Densities having Singularities at the Origin," Communications in Statistics, B, 10, (1981), pp.265-280.

[28] Menke,W.W. "Determination of Warranty Reserves," Management Science, 15, 10, (1969),pp.542-549.

[29] Miller,R.G. and Halpern,J. "Regression with Censored Data," Biomrtrika, 69, (1982), pp.521-531.

[30] Mitra,A. and Patankar,J.G. "Warranty Cost Estimation: A Goal Programming Approach," Decision Sciences, 19, 2, (1988), pp.409-423.

[31] Nguyen,D.G. and Murthy,D.N. "Cost Analysis of Warranty Policies," Naval Research Logistics Quarterly, 31, (1984), pp.525-541.

[32] Patankar,J.G. and Worm,G.H. "Prediction Intervals for Warranty Reserves and Cash Flows," Management Science, 27, 2, (1981),pp.237-241.

[33] Ritchken,P.H. and Tapiero,C.S. "Warranty Design Under Buyers and Seller Risk Aversion," Naval Research Logistics Quarterly, 33,(1986), pp.657-671.

[34] Schneider,H. Lin,B.S. and O'Cinneide,C. "Comparison of Nonparametric Estimators for the Renewal Function," forthcoming in AppliedStatistics, (1990).

[35] Schneider,H. and Weissfeld,l. "Estimation in Linear Models with Censored Data," Biometrika, 73, (1986), pp.741-745.

[36] Shimp,T.A. and Bearden,W.O. "Warranty and Other Extrinsic Cue Effects on Consumers' Risk Perceptions," Journal of Consumer Research, 9, (1982), pp.38-46.

[37] Smith,W.L and Leadbetter,M.R. "On the Renewal Function for the Weibull Distribution," Technometrics, 5, (1963),pp.393-396.

[38] Soland,R.M. "A Renewal Theoretic Approach to the Estimation of Future Demand for Replacement Parts," Operations Research, 16, (1968), pp.36-51.

[39] Soland,R.M. "Availability of Renewal Functions for Gamma and Weibull Distribution With Increasing Hazard Rate," OperationsResearch, 17, (1969), pp.536-543.

[40] Spence,M. "Consumer Misperceptions, Product Failure and Producer Liability," Review of Economic Studies, 44, (1977), pp.561-572.

[41] Sprague,R.H. "A Framework for the Development of Decision Support Systems," MIS Quarterly, 4, 4, (1980), pp.1-26.

[42] Sprague, R.H. and Carlson, E.D. Building Effective Decision Support Systems, New Jersey, Prentice Hall, 1982.

[43] Susarla, V. Tsai, W.Y. and Van Ryzin, J. "A Buckley-James-Type Estimator for the Mean With Censored Data," Biometrika, 71, (1984), pp.624-625.

[44] Udell, J.G. and Anderson, E.E. "The Product Warranty As an Element of Competitive Strategy," Journal of Marketing, 32, (1968), pp.1-8.

PARAMETRIC CERs FOR REPLENISHMENT REPAIR PARTS (Selected Army Air and Ground Systems)

Richard A. Katz

McDonnell Douglas Helicopter Company
5000 E. McDowell Road
Mesa, Arizona 85205-9797

PURPOSE

This research was undertaken during the author's previous employment with CALIBRE Systems, Inc. (CALIBRE) in support of the U.S. Army Cost and Economic Analysis Center (USACEAC). The objective was to assess the feasibility of developing cost estimating relationships (CERs) based on data from the Army Operating and Support Management Information System (OSMIS). The long-range objective is to develop methods to determine total operating and support (O&S) costs within life-cycle cost analysis.

The focus of this effort is on replenishment repair parts cost for selected U.S. Army air and ground systems. Replenishment repair parts are a key element in life-cycle cost, and after military personnel, they are one of the largest components of annual operating cost for a system. Six helicopters and five combat vehicles were selected for this analysis because their replenishment repair parts cost data were available for most active duty divisions in all fiscal years of interest. Requests for a copy of data used in this analysis can be made to Mr. John Pulice, USACEAC, (202) 475-2138 or Autovon 335-2138.

BACKGROUND

Replenishment repair parts are individual parts, assemblies or subassemblies in Army Supply Class IX that are consumable (e.g., switch) or repairable (e.g., gunner's auxiliary sight) below depot level. They are Supply Class IX items identified by a numeric value in the second position of their Materiel Category codes in the Army Master Data File. These parts sustain end items of equipment after fielding. They resupply or replace initial repair parts stocks. This research does not include either initial or war reserves repair parts.

Repair parts costs in this study are based on wholesale requisitions collected from major supply points which fill unit (retail) orders and restock from Army inventories. Funding for the purchase of replenishment repair parts from the Army Operating and Maintenance (OMA) Appropriation is the responsibility of each operating Major Army Command (MACOM).

Replenishment spares, which are funded by the Army procurement appropriations, are not included in this analysis. However, it must be noted that certain shifts in definition between replenishment repair parts and replenishment spares occur from year to year. That is, a certain item classified one year as a spare could be classified in another year as a repair part, and vice versa. All data in this report are based on the Army classification of repair parts and spares which was in effect during FY 87.

This effort is related to the Visibility and Management of Operating and Support Costs (VAMOSC) initiative, which directed the military services to develop automated data systems which would capture and report historical O&S costs of fielded major defense systems. The objective of VAMOSC is to improve O&S cost estimates of new defense systems. It is also intended to provide critical maintenance and support information to promote cost-conscious design of new and fielded systems.

OSMIS is the Army response to VAMOSC. OSMIS reports selected historical O&S data elements for major U.S. Army materiel systems. CALIBRE, under contract to USACEAC, manages OSMIS operation.

DATA COLLECTION AND NORMALIZATION

Costs are identified by MACOM, which for this study include:

- U.S. Army Forces Command (FORSCOM),
- U.S. Army, Europe (USAREUR),
- Eighth U.S. Army (EUSA), and
- U.S. Army Western Command (WESTCOM).

The following total annual data by active duty U.S. Army division for FY 84-87 were extracted from OSMIS:

- replenishment repair parts cost,
- mileage,
- flying hours, and
- density (quantity).

Normalization was required to adjust for an accounting system change, changes in classification between spares and repair parts, and for inflation. Prior to FY 86, OSMIS data were based on shipments. Starting in FY 86, obligation data were collected. To adjust for the accounting change, cost data for FY 84-85 were increased for comparability with obligations. Additionally, FY 84-85 data were normalized to the classification of spares and repair parts which was in effect during FY 87. Parts costs for FY 84-85 were increased by 30 percent based on a ratio of costs before and after the accounting system change (correcting for changes in density and activity). FY 86 and FY 87 data were produced based on obligation data using the FY 87 classification; therefore, no further adjustments were needed. Cost data were escalated to FY 88 constant dollars using the Operations and Maintenance Army (OMA) index.

METHODOLOGY

Number Cruncher Statistical System (NCSS) was used for this effort. Data were analyzed at two levels of detail. One level was an Army-wide perspective which was an aggregation of all division-level data for each system. These data were used to develop CERs related to the life-cycle cost of new systems which the Army might consider for acquisition. The second level consisted of division-level data. These data were used to develop CERs related to allocation of replenishment repair parts funding for systems already fielded. A separate analysis was performed for both new systems and fielded systems.

Army-wide totals by system were investigated for consistency and accuracy. There were several questions to resolve in dealing with these data. Are data consistent with intuition? How are data distributed? Are corrections of erroneous data needed? Are there outliers which should be removed? Is it reasonable to pool data across the four available fiscal years? Scatter plots, box plots, and a test of structural stability were used to provide answers to these questions.

Conventional measures of regression were examined, and methods of Draper and Smith (1981) were used to examine regression residuals. Plots of regression residuals against each explanatory variable were used to search for indications of:

- outliers,
- omitted variables,
- need for data transformation, and
- violations of assumptions of the linear regression model.

Many CERs were developed so that analysts can use whichever parameters are known at the time of the estimate. Numerous variables were examined; however, only relationships which seem intuitively reasonable were retained. All CERs are statistically significant (F test) at 99 percent confidence. CERs were developed using both simple and multiple regression. Both linear and log-log forms were evaluated. CERS were developed for both new systems (based on Army-wide data) and fielded systems (based on division-level data).

The following analysis is divided into two sections--one for helicopters and another for combat vehicles.

HELICOPTERS

The following helicopters were included in this analysis:

- OH-58A (KIOWA) observation helicopter,
- OH-58C (KIOWA) observation helicopter,
- AH-1S (COBRA) attack helicopter,
- UH-1H (IROQUOIS) utility helicopter,
- UH-60A (BLACK HAWK) utility helicopter, and
- CH-47D (CHINOOK) cargo transport helicopter.

These systems were considered to have reached steady state. That is, these systems have been fielded for long enough that their replenishment repair parts costs seem to have stabilized.

As a perspective of helicopters selected for this analysis, Exhibit 1 represents each helicopter using a technique developed by Chernoff (1973). Chernoff Faces are multidimensional graphs in which facial features portray variables over the ranges shown. This seems an effective graphical technique, since people are naturally adept at discriminating among different faces. The six helicopters are grouped by mission. On the left, the lighter OH-58 observation helicopters work in tandem with AH-1S attack helicopters. On the right are light (UH-1H), medium (UH-60A), and heavy (CH-47D) lift helicopters.

For this example, we used only 6 of the 17 facial features NCSS provides. Darkness of the eyes represents the relative

CHERNOFF FACES: HELICOPTER PARAMETERS

FACIAL FEATURE	PARAMETER	RANGE
EYES (IRIS SIZE)	DENSITY	48 - 351
NOSE WIDTH	EMPTY WEIGHT	1464 - 22452 POUNDS
BROW SLANT	MAXIMUM SPEED	127 - 184 MPH
EAR HEIGHT	COST/FH (FY87)	$94 - $409
MOUTH LENGTH	OPTEMPO (FY87)	13 - 17 HOURS/MONTH
MOUTH OPENING	FUEL CONSUMPTION	27 - 400 GAL/HR

Exhibit 1

CALIBRE Systems

quantity of each helicopter in the active-duty Army inventory in FY 87. The UH-1H has the highest density. Nose width indicates relative weight, ranging from light observation helicopters to the heavyweight CH-47D. The more slanted upward the eyebrows, the higher the speed. The AH-1S has the highest speed of the six. Elevation of the ears measures cost per flying hour in FY 87, showing the CH-47D as the most expensive to operate per flying hour. The length of the mouth indicates average monthly flying hours per helicopter in FY 87. For example, the OH-58A was flown fewer hours per month compared to its replacement, the OH-58C. The mouth opening coincides with fuel consumption. The twin engine CH-47D is the largest fuel consumer.

A wide variety of technical parameters were investigated as potential cost drivers of replenishment repair parts cost. A list of these parameters is given in Appendix A. Sources of data were:

- OSMIS,
- Jane's All the World's Aircraft,
- U.S. Military Aircraft Data Book,
- Standard Army Aircraft, and
- U.S. Army Aviation Planning Manual.

DATA ANALYSIS AND CERS (NEW HELICOPTER SYSTEMS)

Exhibit 2 is a summary of Army-wide data in terms of total annual flying hours and total replenishment repair parts cost by helicopter. Both flying hours and parts cost are plotted on the same axes for convenience; however, scaling factors differ. Flying hours are shown in ten thousands, and parts costs are shown in millions. For example, the first set of points for the OH-58A in FY 84 represents approximately $6 million in parts cost and approximately 80,000 flying hours. This chart indicates that OSMIS data seem to make sense over the four years studied. Opposite trends in flying hours reflect the phase-out of the OH-58A and phase-in of the OH-58C. The upward trend for the AH-1S may be a sign of aging, which increases costs. The UH-1H pattern seems stable. The UH-60A fleet was grounded for part of FY 85, but replenishment repair parts cost was incurred for deferred maintenance. However, fewer hours were flown, distorting the cost per flying hour. This accounts for the relatively large gap between flying hours and replenishment repair parts cost in

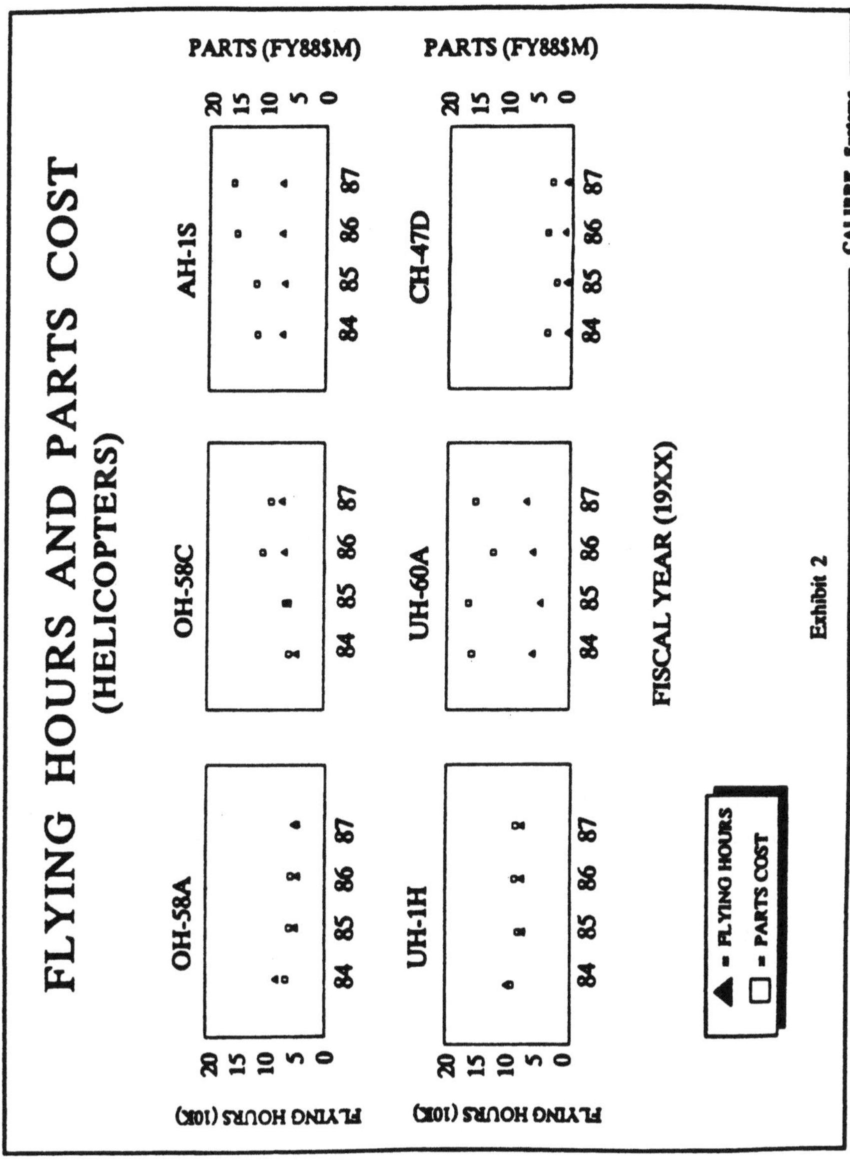
FLYING HOURS AND PARTS COST
(HELICOPTERS)
OH-58A
OH-58C
AH-1S
UH-1H
UH-60A
CH-47D
FLYING HOURS (10K)
PARTS (FY88$M)
20 15 10 5 0
84 85 86 87
FISCAL YEAR (19XX)
= FLYING HOURS
= PARTS COST
CALIBRE Systems

Exhibit 2

that year. We understand there may have been a similar condition for the CH-47D in FY 84. Also, the relatively large gap between flying hours and parts cost for the CH-47D in FY 84 may indicate a newly fielded system which had not yet reached steady state. Conventionally, O&S cost estimation within life-cycle cost analysis is based on steady-state analysis; therefore, we decided to remove these two points (FY 85 UH-60A and FY 84 CH-47D) from further analysis.

For development of CERS, 22 observations were used: four years of data for six helicopters, less the two points removed. Both linear and logarithmic forms were employed. CERs for new systems were based on independent variables which drive Army-wide cost per flying hour. A summary of CERs is shown in Exhibit 3 in order of R^2. The percent error shown is the ratio of the root mean square error to the mean of the response variable. In the linear CERs, this ratio is the Coefficient of Variation. In nonlinear CERs, this ratio is ± antilog of the root mean square error MINUS one.

The strongest cost drivers in linear forms were those expected: weight, engine horsepower, fuel consumption, and fuel capacity. These variables have appeared in previous CER research of aircraft.

In the log-log set of equations, an interesting CER is the relation of cost per flying hour to approximately the square of maximum speed. This may be related to a theory of mechanics that drag increases by the square of speed. Another interesting finding is the relative weakness of maximum speed as a cost driver in this analysis. This fact is consistent with speed limitations for conventional helicopters. Gablehouse (1969) points out that:

> Since the lift on the blades on the advancing side cannot be permitted to exceed the lift on the retreating side, this factor tends to limit the top speed at which the fastest helicopters (with conventional rotor systems) can fly; for even the largest and most powerfully-engined of helicopters, at speeds just above 200 miles per hour the lifting and propelling characteristics of the rotor are affected, and a phenomenon termed "blade stall" is encountered.

Various multiple regression relationships were investigated. Some combinations of independent variable were less successful because of high intercorrelation among them. However, year of first production proved useful because it

CER EQUATIONS (FY88$)
NEW HELICOPTER SYSTEMS

LINEAR FORMS	R^2 (ADJ)	PERCENT ERROR
COST/FH = 90.09 + .012 (EMPTY WEIGHT)	.86	19%
COST/FH = 101.50 + .035 (SHAFT HORSEPOWER)	.86	19%
COST/FH = 98.41 + .68 (FUEL CONSUMPTION)	.84	20%
COST/FH = 99.29 + .040 (FUEL CAPACITY)	.83	21%
COST/FH = 105.69 + .0053 (TAKEOFF WEIGHT)	.82	22%
COST/FH = 123.87 + .011 (PAYLOAD)	.76	25%
COST/FH = 85.44 + 53.77 (PRICE)	.70	27%
COST/FH = 70.64 + .011 (ENGINE RPM)	.49	36%

Exhibit 3

CALIBRE Systems

CER EQUATIONS (FY88$)

NEW HELICOPTER SYSTEMS (CONTINUED)

LOG-LOG FORMS	R^2 (ADJ)	PERCENT ERROR
COST/FH = .00029 (CRUISE SPEED)$^{2.69}$	.75	26%
COST/FH = 2.56 (MAIN ROTOR HEIGHT)$^{1.67}$	.68	29%
COST/FH = 1.14 (OVERALL LENGTH)$^{1.24}$	.62	32%
COST/FH = .0014 (RATE OF CLIMB)$^{1.54}$	.62	33%
COST/FH = .0047 (MAXIMUM SPEED)$^{2.07}$	.61	33%
COST/FH = .14 (FUSELAGE LENGTH)$^{1.90}$	.57	35%
COST/FH = .0000049 (YEAR OF 1ST PRODUCTION)$^{4.01}$	.54	36%
COST/FH = .27 (MAIN ROTOR DIAMETER)$^{1.68}$	.52	37%
COST/FH = .15 (HOVERING CEILING OUT-OF-GROUND EFFECT)$^{.77}$	.45	40%

Exhibit 3 (Cont'd)

CALIBRE Systems

CER EQUATIONS (FY88$)

NEW HELICOPTER SYSTEMS (CONTINUED)

MULTIVARIATE	R^2 (ADJ)	PERCENT ERROR
COST/FH = .00029 (FUEL CONSUMPTION) $^{.31}$ *(YEAR OF 1ST PRODUCTION) $^{2.75}$	.90	16%
COST/FH = .00015(EMPTY WEIGHT) $^{.30}$ *(YEAR OF 1ST PRODUCTION) $^{2.63}$	.88	17%
COST/FH = 229.64 - 95.44 (M1) + 132.61 (M2)	.81	22%

DUMMY VARIABLES M1 AND M2 REPRESENT AIRFRAME MANUFACTURERS: BELL (1,0); BOEING (0,1); SIKORSKY (0,0).

Exhibit 3 (Cont'd)

CALIBRE Systems

added explanatory power but was not highly correlated with other independent variables. Year of first production can serve as a measure of technological level. Two CERs incorporating year of first production in conjunction with other variables are shown as examples of this form.

Several dummy variables were considered; however, only airframe manufacturer was a useful dummy variable in explaining cost per flying hour. Dummy variables for engine manufacturer, MACOM, and unit type were tested but did not prove significant cost drivers. It is not surprising that the smaller, lighter Bell helicopters (AH-1S, OH-58A/C, and UH-1H) have a lower cost per flying hour than the Boeing CH-47D or the Sikorsky UH-60A. Dummy variables for manufacturer are, in essence, a substitute for weight, which already proved a strong cost driver.

In the course of regression analysis and analysis of variance, we used a statistical test usually attributed to Chow (1960) to determine whether or not data should be treated individually for each year or pooled over four fiscal years. Based on a test of all variables used in linear CERs at 95 percent confidence, we found the regression coefficient remained stable over the four years. Therefore, it was not unreasonable to pool data for all four fiscal years.

Once CERs were developed and tested, the next step was to use them for prediction. As an example, Exhibit 4 shows a prediction for the LHX helicopter, which the Army is considering for acquisition. The prediction is based on the CER for empty weight as a driver of parts cost per flying hour. The regression, line, 95 percent confidence interval (mean), and data points are shown. Based on the projected empty weight of 7,500 pounds, predicted parts cost per flying hour for the LHX is \$181 ± \$15. This estimate applies to the steady state (when cost per flying hour stabilizes some time after initial fielding).

DATA ANALYSIS AND CERS (FIELDED HELICOPTER SYSTEMS)

Division-level data were investigated to develop CERs for fielded systems. After preliminary evaluation of the data, we decided to remove outliers conservatively. Using methods suggested by Chatterjee and Hadi (1986), outliers were identified using studentized residuals from regression of parts cost on density, flying hours, and empty weight. Studentized residuals are those scaled by the standard error based on a

CER APPLICATION: LHX PREDICTION
(95% CONFIDENCE LEVEL--MEAN)

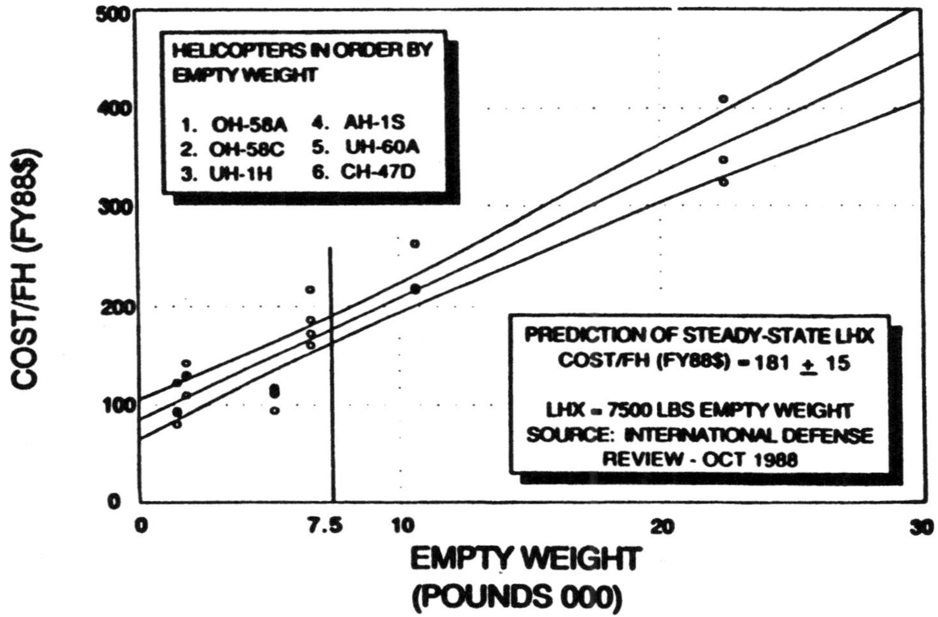

Exhibit 4

CALIBRE Systems

regression plane fitted to all other points. A calibration point or cutoff value of 2.5 was chosen for the absolute value of studentized residuals. This corresponds to a 99 percent confidence band for a t-distribution. Observations whose studentized residuals exceeded the cutoff value were considered outliers and removed. In this process, we removed 26 of 290 data points from analysis of fielded systems; no change was made to the data base for new systems.

Box plots (Tukey, 1977) were used to provide insight into division-level data. Exhibit 5 contains box plots of parts cost per flying hour and parts cost per helicopter for FY 84-87 by type of helicopter in order of increasing empty weight. The fact that these parts cost rates increase with weight reinforces the earlier finding that weight is a relevant cost driver.

Each box contains data points for all units for FY 84-87. The middle line in the box plot is the median of the data. For example, the median cost per flying hour for the AH-1S is approximately $200. The top and bottom of the box are boundaries for the middle half of the data. The whiskers (vertical lines emanating from the box) terminate at the 10th and 90th percentiles of the data. Values which appear beyond the whiskers are potential outliers, such as units with cost per helicopter (second page of Exhibit 5) above $30,000 for the OH-58A. In keeping with a conservative approach, we decided not to remove any more outliers.

Linear regression was used as a tool to develop CERs for fielded systems. As a first step, flying hours and density were examined as potential cost drivers. Based on previous results, we continued to pool data for all four fiscal years.

A summary of CERs is shown in Exhibit 6. CERs based on division-level data provided an unexpected result. We had expected flying hours to be strongly related to repair parts cost. Instead, we found that flying hours was not as strong as density as a cost predictor. Possible causes of this phenomenon are policy and fiscal limitations which affect flying hour allocations.

Useful rules of thumb emerged in terms of both cost per flying hour and cost per helicopter. In the first equation, the cost per flying hour for the average helicopter is $150. Based on the second equation, each helicopter adds approximately $21,000 in replenishment repair parts cost. These values are useful in the absence of more sophisticated methods. This situation can be improved by addition of a technical characteristic along with density. In the third equation, empty weight proves a useful addition to density. As an alternative to empty weight, use of dummy variables for airframe manufacturer (which are

BOX PLOTS (HELICOPTERS)

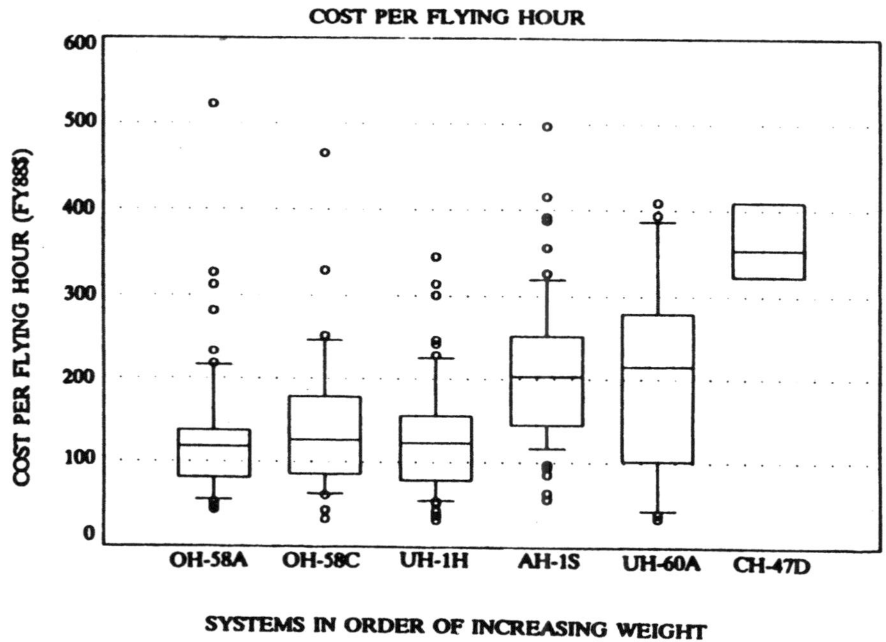

Exhibit 5

BOX PLOTS (HELICOPTERS) (CONT'D)

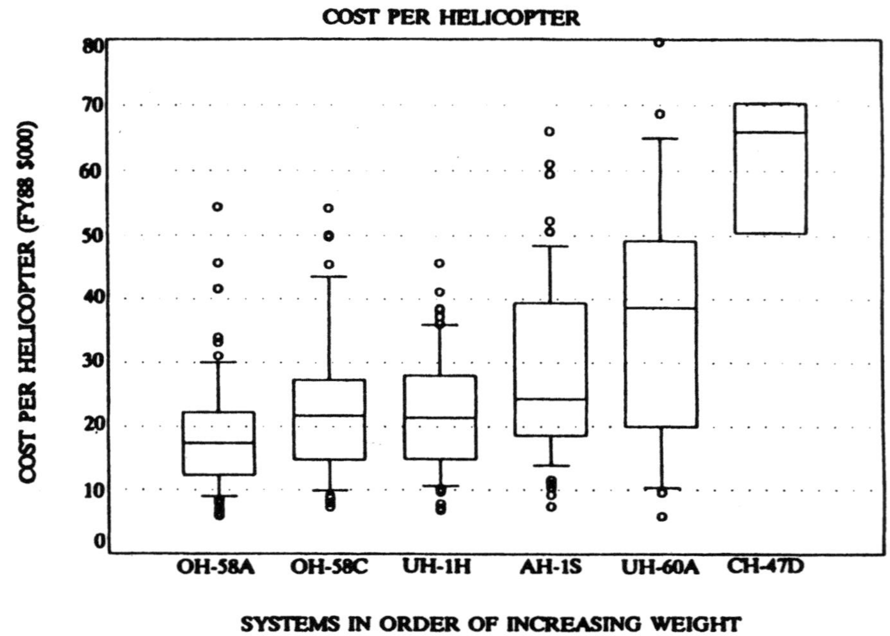

Exhibit 5 (Cont'd)

CALIBRE Systems

CER EQUATIONS (FY88$M)
FIELDED HELICOPTER SYSTEMS

LOG-LOG FORMS	R^2(ADJ)	PERCENT ERROR
ANNUAL REPAIR PARTS COST = $.00015\,(\text{FLYING HOURS})^{.99}$	.64	82%
ANNUAL REPAIR PARTS COST = $.021\,(\text{DENSITY})^{1.0}$	.70	73%
ANNUAL REPAIR PARTS COST = $.0025\,(\text{DENSITY})^{.97}$ *$(\text{EMPTY WEIGHT})^{.28}$	.74	66%
DUMMY VARIABLE FORM		
ANNUAL REPAIR PARTS COST = .49 + .029 (DENSITY) - .66 (M1) + 1.25 (M2)	.72	53%

* DUMMY VARIABLES M1 AND M2 REPRESENT AIRFRAME MANUFACTURERS: BELL (1,0); BOEING (0,1); SIKORSKY (0,0).

Exhibit 6

CALIBRE Systems

a substitute for weight) also improves the situation. The error percentages for these CERs seem large, because there is more variability in dealing with data for individual divisions as opposed to Army-wide aggregates of these data.

COMBAT VEHICLES

The following combat vehicles were included in this research:

- M1 (Abrams) main battle tank,
- M60A3 main battle tank,
- M88A1 medium recovery vehicle,
- M2/M3 Bradley Fighting Vehicle, and
- M109A3 self-propelled howitzer.

As a perspective of combat vehicles selected for this analysis, Exhibit 7 represents each system as a Chernoff Face with features similar to those used for helicopters. The five combat vehicles are grouped by mission. On the left are heavy tanks and recovery vehicle. On the right are the Bradley Fighting Vehicle and a self-propelled howitzer. The M1 (darkness of the eyes) has the highest density. The M60A3 is slower (eyebrow slant) but cheaper to operate (elevation of ears) than the M1. Because of its mission profile, we would expect annual mileage (optempo), represented by mouth length, to be low for the M88A1. The Bradley Fighting Vehicle and M109A3 are light vehicles (nose width) which are relatively inexpensive to operate (elevation of ears).
A list of technical parameters which were investigated as potential cost drivers for replenishment repair parts cost are shown in Appendix B. The sources of these data were:

- OSMIS, and
- Jane's Armour and Artillery.

DATA ANALYSIS AND CERS (NEW COMBAT VEHICLE SYSTEMS)

Exhibit 8 is a summary of Army-wide data in terms of total annual mileage and total replenishment repair parts cost by combat vehicle. Both mileage and parts cost are plotted on the same axes for convenience; however, scaling factors differ. Mileage is shown in hundred thousands, and parts costs are shown in millions. For example, the first set of points for the M1 tank in FY 84 represents approximately $30 million in

CHERNOFF FACES: COMBAT VEHICLE PARAMETERS

M1 TANK	M60A3 TANK	M88A1 RECOVERY	M2/M3 BRADLEY	M109A3 HOWITZER

FACIAL FEATURE	PARAMETER	RANGE
EYES (IRIS SIZE)	DENSITY *	593 - 1293
NOSE WIDTH	COMBAT WEIGHT	25 - 60 TONS
BROW SLANT	MAXIMUM SPEED	26 - 45 MPH
EAR HEIGHT	COST/MILE (FY88) *	$15 - $33
MOUTH LENGTH	ANNUAL OPTEMPO *	651 - 910 MILES
MOUTH OPENING	FUEL CONSUMPTION	1 - 7 GAL/MILE

* FY87 SNAPSHOT

Exhibit 7

CALIBRE Systems

MILES AND PARTS COST
(COMBAT VEHICLES)

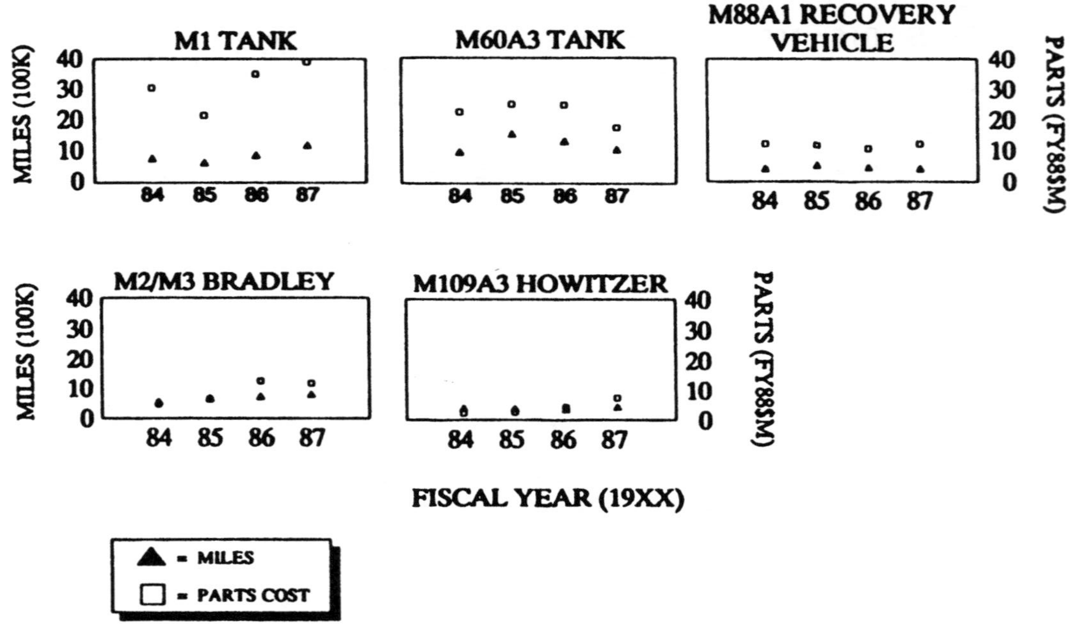

Exhibit 8

CALIBRE Systems

parts cost and approximately 800,000 miles driven. This chart indicates that OSMIS data seem to make sense over the four years studied. The decrease in total parts cost for the M1 in FY 85 is the result of a decrease in optempo. The increase in M1 cost and mileage starting in FY 86 results from an increase in systems fielded. The decreasing trend in M60A3 values reflects the replacement of this system with the M1A1 tank. The M881A values remain stable over the four years examined. The increasing trend in the Bradley Fighting Vehicle data is due to an increase in systems fielded. The level for the M109A3 is fairly stable.

For development of CERs, 20 observations were used: four years of data for five combat vehicles. CERs were developed using simple and multiple linear regression. Both linear and logarithmic forms were employed. A summary of CERs is shown in Exhibit 9.

The strongest cost drivers in linear forms were engine horsepower and fuel capacity, as might be expected. Both equations were developed with a zero intercept.

For other cost drivers, parts cost does not appear a linear function of mileage. Therefore, instead of using a cost rate (such as cost per mile or cost per vehicle), parts cost was used as the response variable. Either mileage or density was used as a cost driver along with technical characteristics. In conjunction with density, engine speed (RPM) proved a particularly strong driver of parts cost. Track length proved strongest in conjunction with mileage. Another CER of interest involves price, which exhibited a nonlinear relationship with parts cost. This does not agree with commonly held practice of determining parts cost as a straight percentage of price. The negative power in the CER based on height seems unusual. However, it is logical since heavy tanks built low to the ground are more expensive to operate than the relatively tall M109A3 howitzer.

We expected weight to be a relatively strong cost driver; however, this was not the case. One reason is that there were only two weight groups represented in this analysis: 56-60 tons (tanks and recovery vehicle) and 25-27 tons (Bradley and howitzer). In the future, a better design would include systems in additional weight classes. For example, including the 12-ton M113A armored personnel carrier would probably improve the explanatory power of weight as a cost driver.

In the course of regression analysis and analysis of variance, we used the Chow procedure to determine whether or not data should be treated individually for each year or pooled over

CER EQUATIONS (FY88$)
NEW COMBAT VEHICLE SYSTEMS

			R^2 (ADJ)	PERCENT ERROR
LINEAR FORMS				
COST/MILE	=	.026 (HORSEPOWER)	.85	18%
COST/MILE	=	.067 (FUEL CAPACITY)	.76	23%
LOG-LOG FORMS				
PARTS COST	=	$17\ (\text{DENSITY})^{1.4}\ (\text{KRPM})^{4.3}$	.90	29%
PARTS COST	=	$.00000012\ (\text{MILES})^{1.2}\ (\text{TRACK LENGTH})^{6.2}$	.88	31%
PARTS COST	=	$.0000000075\ (\text{MILES})^{1.5}\ (\text{HULL LENGTH})^{4.8}$	.87	34%
PARTS COST	=	$9700\ (\text{MILES})^{.54}\ (\text{WIDTH-10})^{.58}$	.83	40%

Exhibit 9

CALIBRE Systems

CER EQUATIONS (FY88$)

NEW COMBAT VEHICLE SYSTEMS (CONTINUED)

LOG-LOG FORMS			R^2 (ADJ)	PERCENT ERROR
PARTS COST	=	$44\,(\text{DENSITY})^{.99}\,(\text{TRENCH CLEARANCE})^{2.8}$	.82	40%
PARTS COST	=	$580\,(\text{DENSITY})^{1.3}$ * $(\text{VERTICAL OBSTACLE CLEARANCE})^{.96}$	.82	41%
PARTS COST	=	$8200\,(\text{DENSITY})^{1.0}\,(\text{FUEL CONSUMPTION})^{.55}$	.78	45%
PARTS COST	=	$45000\,(\text{DENSITY})^{.83}\,(\text{PRICE})^{1.5}$	.77	46%
PARTS COST	=	$6600\,(\text{DENSITY})^{.98}\,(\text{TRACK WIDTH})^{1.6}$	.75	49%
PARTS COST	=	$670\,(\text{DENSITY})^{.94}\,(\text{COMBAT WEIGHT})^{.96}$	.75	50%
PARTS COST	=	$.13\,(\text{MILES})^{1.4}\,(\text{HEIGHT-9})^{-.64}$	.75	50%
PARTS COST	=	$.000024\,(\text{MILES})^{.77}\,(\text{RANGE})^{3.0}$	.65	60%

Exhibit 9 (Cont'd)

CALIBRE Systems

CER EQUATIONS (FY88$)

NEW COMBAT VEHICLE SYSTEMS (CONCLUDED)

LOG-LOG FORMS	R^2 (ADJ)	PERCENT ERROR
PARTS COST = 12000 $(\text{MILES})^{1.8}$ $(\text{FIRST PRODUCTION})^{-5.1}$	.62	64%
PARTS COST = .012 $(\text{MILES})^{1.3}$ * $(\text{POWER-TO-WEIGHT RATIO})^{1.07}$	.63	63%
PARTS COST = 170 $(\text{DENSITY})^{1.4}$ $(\text{MAXIMUM SPEED})^{.45}$	.59	67%

Exhibit 9 (Cont'd)

CALIBRE Systems

four fiscal years. At 95 percent confidence, the CERs for both engine horsepower and fuel capacity were stable over the four years. Based on these findings, we decided to continue pooling data for all four fiscal years.

Once CERs were developed and tested, the next step was prediction. As an example, Exhibit 10 shows a prediction for the Fire Support Team Vehicle (FISTV), a variant of the armored personnel carrier. The FISTV was only recently fielded, so it has not yet had time to reach steady state in operating cost. The prediction for the FISTV is based on the CER for engine horsepower as a driver of parts cost per mile. The regression line, 95 percent confidence interval (mean), and data points are shown. Based on the FISTV engine design of 215 horsepower, predicted cost per mile is \$6.60 ± \$3.00. This estimate applies to the steady state.

DATA ANALYSIS AND CERS (FIELDED COMBAT VEHICLE SYSTEMS)

Divison-level data were investigated to develop CERs applicable to systems already fielded. As with helicopters, outliers were removed on the basis of studentized residuals. In this case, 5 outliers were removed, leaving 172 data points for this analysis.

Box plots were used to provide insight into division-level data. Exhibit 11 contains box plots of parts cost per mile and parts cost per combat vehicle for FY 84-87 in order of increasing engine horsepower. The fact that these parts cost rates increase with engine horsepower reinforces the earlier finding that engine horsepower is a relevant cost driver.

Linear regression was used as a tool to develop CERs for fielded systems. In addition to technical parameters used for Army-wide analysis, mileage and density were examined as potential cost drivers.

Useful CERs were developed for replenishment repair parts based on both mileage and density. A summary of CERs is shown in Exhibit 12. Since record keeping for mileage is difficult, we had expected density to be a stronger cost driver than mileage. We were surprised to find the opposite was true. Next we found that CERs based on simple regressions of parts cost on either mileage or density could be improved by addition of a technical characteristic. For example, engine horsepower proved a useful addition to density in explaining parts cost.

CER APPLICATION: FISTV PREDICTION
(95% CONFIDENCE LEVEL--MEAN)

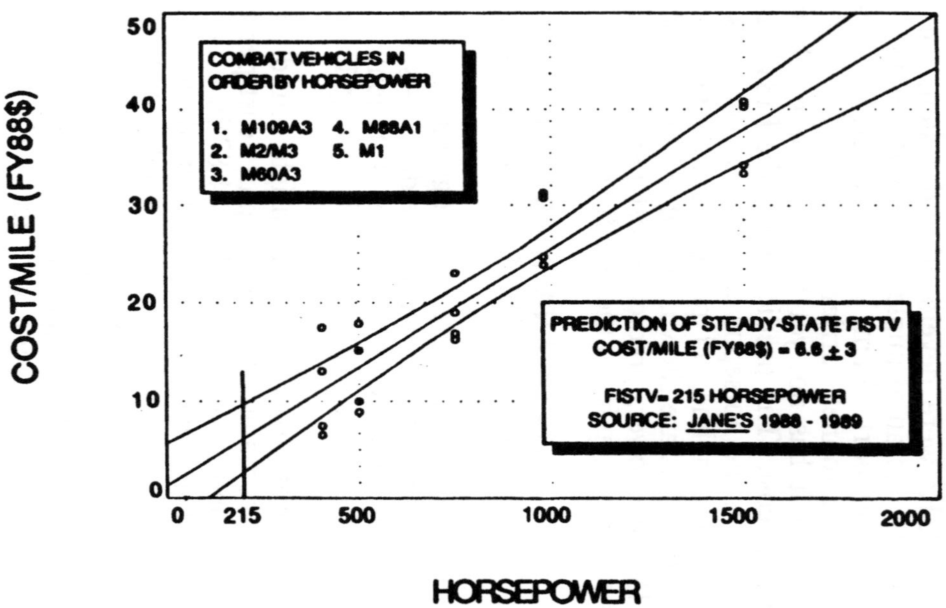

Exhibit 10

CALIBRE Systems

BOX PLOTS (COMBAT VEHICLES)

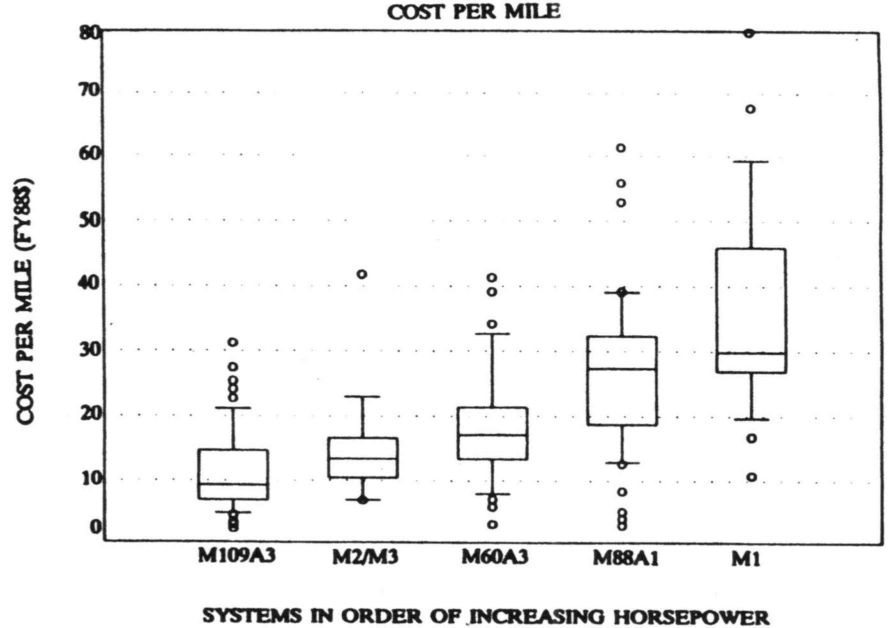

Exhibit 11

CALIBRE Systems

BOX PLOTS (COMBAT VEHICLES) (CONT'D)

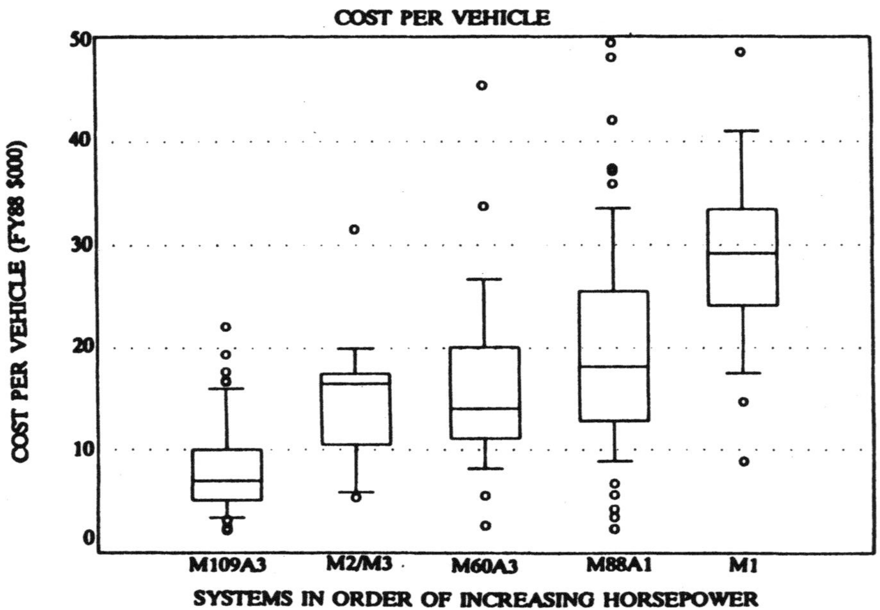

Exhibit 11 (Cont'd)

CER EQUATIONS (FY88$)
FIELDED COMBAT VEHICLE SYSTEMS

LOG-LOG FORMS	R^2 (ADJ)	PERCENT ERROR
ANNUAL REPAIR PARTS COST = 7.5 $(\text{MILES})^{1.1}$	.81	86%
ANNUAL REPAIR PARTS COST = 7000 $(\text{DENSITY})^{1.2}$	.75	99%
ANNUAL REPAIR PARTS COST = .037 $(\text{MILES})^{1.0}$ * $(\text{HORSEPOWER})^{.87}$	.88	62%
ANNUAL REPAIR PARTS COST = 16 $(\text{DENSITY})^{1.1}$ * $(\text{POWER-TO-WEIGHT RATIO})^{.90}$ * $(\text{FUEL CAPACITY})^{.69}$	.91	27%

Exhibit 12

RESEARCH BENEFITS

This research proved that OSMIS is an excellent source of data for development of CERs for both air and ground systems. While there is still room for improvement, the data were consistent and well behaved. OSMIS is a mature data base ready to support increased analysis of U.S. Army O&S costs.

CERs developed in this research are statistically significant; they are also consistent with intuition. These models can be used to support Army decisions concerning costs of both new and fielded systems. This research is a beginning. There is certainly room for further investigation into every relationship developed here. These CERs should be tested continually against operational reality. Refinement and replacement of CERs should be a continuing research effort. USACEAC is already in the process of validating and extending this analysis.

This effort lays important groundwork for improved O&S analysis. Methodologies which were developed can be used to extend the analysis to additional fiscal years and additional systems. Furthermore, there is a potential to broaden this research into other O&S cost areas which OSMIS contains, such as depot maintenance, POL, and ammunition. These procedures can support a complete O&S model.

This research is a landmark event for Army VAMOSC. Is shows OSMIS to be a source of believable, reliable O&S data. It demonstrates the feasibility of using OSMIS data to develop useful tools for O&S cost estimating. It also shows USACEAC commitment to release of accurate, validated data under its VAMOSC charter.

BIBLIOGRAPHY

Chatterjee, Samprit and Ali S. Hadi, "Influential Observations, High Leverage Points, and Outliers in Linear Regression," Statistical Science, Vol. 1, No. 3, August 1986, pp. 379-416.

Chernoff, Herman, "The Use of Faces to Represent Points in k-Dimensional Space Graphically," Journal of the American Statistical Association, Vol. 68, No. 342, June 1973, pp. 361-368.

Chow, G., "Tests for Equality Between Sets of Coefficients in Two Linear Regressions," Econometrica, Vol. 28, July 1960, pp. 591-605.

Department of the Army Field Manual 101-20, U.S. Army Aviation Planning Manual, 6 January 1984.

Draper, Norman R. and Harry Smith, Jr., Applied Regression Analysis, Second Edition, John Wiley & Sons, Inc., New York, 1981.

Foss, Christopher F. (ed.), Jane's Armour and Artillery, 1988-9, Jane's Publishing Inc., New York, 1986.

Gablehouse, Charles, Helicopters and Autogiros, J.B. Lippincott, Philadelphia, 1969.

Hintze, Jerry L., Number Cruncher Statistical System, Version 5.1, Kaysville, Utah, 1987.

Katz, Richard A., "Parametric Replenishment Repair Parts Analysis," Proceedings of the International Society of Parametric Analysts, May 1989, pp. 387-416.

Katz, Richard A. and John E. Berterman, Geographical Differences in Replenishment Repair Parts Costs for Selected U.S. Army Helicopters, WN-8815-6, CALIBRE Systems, Inc., 19 April 1989.

_______, Cost Estimating Relationships for Replenishment Repair Parts of Selected U.S. Army Helicopters, WN-8715-10, Management Consulting & Research, Inc., 1 September 1988.

Nicholas, Ted G., U.S. Military Aircraft Data Book, 1987, Data Research Associates, Fountain Valley, California, 1986.

Standard Army Aircraft, U.S. Army Aviation Center, Fort Rucker, Alabama, undated.

Taylor, John W. R. (ed.), Jane's All the World's Aircraft, 1986-7, Jane's Publishing Inc., New York, 1986.

Tukey, John W., Exploratory Data Analysis, Addison-Wesley Publishing Company, Reading, Massachusetts, 1977.

APPENDIX A: VALUE OF TECHNICAL CHARACTERISTICS BY HELICOPTER

PHYSICAL PARAMETERS

	MAIN ROTOR DIAM (FT)	MAIN ROTOR HEIGHT (FT)	FUSELAGE LENGTH (FT)	OVERALL LENGTH (FT)	EMPTY WEIGHT (LBS)	PAYLOAD (LBS)	FUEL CAPACITY (LBS)
UH-60A	53.7	12.3	50.0	64.8	10624	8000	2198
OH-58C	35.3	9.5	32.6	41.0	1818	284	456
AH-1S	44.0	13.4	44.6	53.1	6598	1096	1795
UH-1H	48.0	11.8	41.9	57.8	5210	3880	1370
OH-58A	35.3	9.5	32.6	41.0	1464	170	455
CH-47D	60.0	18.7	51.0	99.0	22452	22000	6700

PERFORMANCE PARAMETERS

	MAX SPEED (MPH)	CRUISE SPEED (MPH)	HOVER CEILING* (FT)	CLIMB RATE (FT/MIN)	ENGINE POWER (HP)	FUEL CONSUM (GAL/HR)
UH-60A	184	167	10400	2460	3120	148
OH-58C	138	117	9700	1800	420	27
AH-1S	195	139	9900	1620	1800	98
UH-1H	127	127	4000	1600	1400	84
OH-58A	138	117	8800	1780	317	29
CH-47D	181	167	15000	3100	7500	400

* OUT OF GROUND EFFECT

OTHER PARAMETERS

	PRICE ($FY80)	YEAR OF 1ST PROD (19XX)
UH-60A	2.28	78
OH-58C	0.20	79
AH-1S	2.48	77
UH-1H	1.76	65
OH-58A	0.14	69
CH-47D	4.28	82

APPENDIX B: VALUE OF TECHNICAL CHARACTERISTICS BY COMBAT VEHICLE

PHYSICAL PARAMETERS

	COMBAT WEIGHT (TONS)	HULL LENGTH (FT)	WIDTH (FT)	HEIGHT (FT)
M60A3	58.00	22.79	11.91	10.73
M1	60.12	25.98	11.98	9.47
M109A3	27.53	22.63	10.33	10.76
M88A1	56.00	27.12	11.25	10.58
M2/M3	24.67	21.16	10.50	9.75

	TRACK WIDTH (FT)	TRACK LENGTH (FT)	FUEL CAPACITY (GAL)
M60A3	2.33	13.89	375
M1	2.08	15.26	504
M109A3	1.25	13.00	135
M88A1	2.33	15.12	400
M2/M3	1.75	12.83	175

PERFORMANCE PARAMETERS

	MAX SPEED (MPH)	ROAD RANGE (MILES)	OBSTACLE CLEARANCE (FT)	TRENCH CLEARANCE (FT)
M60A3	30	298	3.00	8.50
M1	45	309	4.08	9.00
M109A3	35	220	1.74	6.00
M88A1	26	280	5.58	8.56
M2/M3	41	300	3.00	8.33

	ENGINE POWER (HP)	ENGINE SPEED (RPM)	FUEL CONSUM (GAL/MI)	POWER-WEIGHT RATIO (HP/TON)
M60A3	750	2400	1.90	12.93
M1	1500	3000	7.29	24.95
M109A3	405	2300	1.18	14.71
M88A1	980	2800	1.46	17.50
M2/M3	500	2600	1.12	20.27

OTHER PARAMETERS

	PRICE (FY89$)	YEAR OF 1ST PROD (19XX)
M60A3	1.183	80
M1	1.646	80
M109A3	0.758	78
M88A1	0.837	76
M2/M3	1.010	81

ACRONYMS

CER	Cost Estimating Relationship
EUSA	Eighth U.S. Army
FISTV	Fire Support Team Vehicle
FORSCOM	U.S. Army Forces Command
MACOM	Major Army Command
NCSS	Number Cruncher Statistical System
O&S	Operating and Support
OMA	Operating and Maintenance, Army
OSMIS	Operating and Support Management Information System
USACEAC	U.S. Army Cost and Economic Analysis Center
USAREUR	U.S. Army, Europe
VAMOSC	Visibility and Management of Operating and Support Costs
WESTCOM	U.S. Army Western Command

VII. Test and Evaluation Issues

Maintaining the Capital Stock at DoD Test and Evaluation Sites

Daniel B. Levine

K. M. Olver

The purpose of this study was to consider the question of whether the military services have been replenishing the capital stock (the value of facilities and equipment) at their test and evaluation (T&E) sites at an adequate rate. Adequacy is determined by comparing the average age and renewal period of the total capital stock at the T&E sites with figures for other sectors of the economy: nine major aerospace firms, major U.S. industries, and the military and civilian components of the federal government. Although aggregate analysis leaves many questions unanswered, the findings show unequivocally that the T&E sites are lagging far behind the other sectors at present, and that the disparity will worsen in time unless the services increase their rate of investment.

Daniel B. Levine
Institute for Defense Analyses
Cost Analysis and Research Division
1801 N. Beauregard Street
Alexandria, Virginia 22311
(703) 845-2562

MAINTAINING THE CAPITAL STOCK AT DoD TEST AND EVALUATION SITES

Daniel B. Levine
Institute for Defense Analyses
1801 N. Beauregard Street
Alexandria, VA

K. M. Olver
Institute for Defense Analyses
1801 N. Beauregard Street
Alexandria, VA

INTRODUCTION

This paper addresses the question of whether the Department of Defense (DoD) has been spending enough to replenish the capital stock at military test and evaluation (T&E) sites. For purposes of this paper, "capital stock" refers to the physical plant comprised of structures and major equipment, which, at a T&E site, would include buildings, towers, test stands, range instrumentation, data processing equipment, etc. We have excluded the implicit costs of land, the third major category of capital. The analysis focuses on the capital stock at the 22 principal T&E sites, those belonging to the Major Range and Test Facility Base (MRTFB), shown in Table 1.

Some recent work within DoD suggests that investment in the T&E sites has, indeed, been too low. A recent analysis by the Director, Operational Test & Evaluation, found that the military services are almost $15 billion short of the new equipment and facilities needed to meet testing requirements through 1994. The analysis cited specific needs in important areas: facilities to test advanced avionics equipment in simulated battlefield conditions, realistic "targets" to test ballistic missile defenses, and hypersonic windtunnels, sensor test chambers, environmental simulators, and nuclear effects simulators to test new-technology weapons.

Table 1. Major Range and Test Facility Base

Service	Sites
Army	Yuma Proving Ground, Yuma, AZ
	Aberdeen Proving Ground, Aberdeen, MD
	Dugway Proving Ground, Salt Lake City, UT
	White Sands Missile Range, White Sands, NM
	Electronic Proving Ground, Fort Huachuca, AZ
	High Energy Laser Systems Test Facilities, White Sands, NM
	Kwajalein Missile Range, Kwajalein Atoll, Marshall Islands
Navy	Atlantic Undersea Test and Evaluation Center, Andros Island
	Naval Air Propulsion Center, Trenton, NJ
	Naval Air Test Center, Patuxent, MD
	Naval Weapons Center, China Lake, CA
	Pacific Missile Test Center, Point Mugu, CA
	Atlantic Fleet Weapons Training Facility, PR
Air Force	4950th Test Wing, Wright Patterson AFB, OH
	6585th Test Group, Holloman AFB, NM
	Armament Development and Test Center, Eglin AFB, FL
	Arnold Engineering Development Center, Arnold AFB, TN
	Air Force Flight Test Center, Edwards AFB, CA
	Eastern Space and Missile Center, Patrick AFB, FL
	Western Space and Missile Center, Vandenberg AFB, CA
	Air Force Tactical Fighter Weapons Center, Nellis AFB, NV
	Utah Test and Training Range, Hill AFB, UT

The present study is intended to shed some light on this issue by comparing the rate of capital renewal at the T&E sites with the rate elsewhere in the U.S. economy: the major U.S. industries, the nine principal aerospace firms, and both the non-military and military components of the federal government.[1]

The private sector is a reasonable standard for comparison because the managers of private firms have profit incentives to maintain capital at efficient levels. It is also relevant to compare the T&E sites with the federal government as a whole: this comparison gives us an indication of how the T&E sites are doing relative to other

1 The sectors being compared are not completely separate: The aerospace firms are a part of the major industries, and the DoD T&E sites are a part of the military component of the federal goverment.

organizations that operate under incentives characteristic of government bureaucracies.

If we find that the T&E sites have been renewing capital at a much slower pace than the private and government sectors as a whole, that would suggest that the T&E sites are falling behind. In that case, a more detailed analysis should be conducted to confirm the overall finding and to identify the specific test capabilities that need improvement.

We measured the rate of capital renewal by two variables. The first is the average age of the capital stock. Aging capital means that buildings are becoming more costly to maintain and, for the particular case of T&E sites, that equipment may be becoming outmoded and less capable of testing new systems.

The second variable is the capital renewal period obtained by dividing the capital stock (the total value of structures and equipment) by the level of investment (the amount that the capital stock increases in one year). Whereas the average age reflects the result of *past* capital replenishment policies, the renewal period indicates the number of years it would take in the *future* to completely replace the capital stock at current levels of investment. The longer the renewal period, the older the stock of buildings and equipment the facility will be operating with in the future.

SUMMARY

The finding of the analysis is that the T&E sites have been renewing capital at a much slower rate than the other sectors. As shown in Table 2, the T&E sites' capital stock is much older, and their renewal period is much longer. Moreover, these two measures together imply that the disparity between the T&E sites and the other sectors will get worse in time. Dividing the capital renewal period by two (2) yields an estimate of the future average age of the capital stock if investment continues at the current rate.[2] The figures in Table 2 indicate that for the sectors other than the DoD T&E sites, the future average age is substantially less than the average age, which suggests that the average age will decline in time. For the DoD T&E sites, however, the future average age is about the same as the current average age, suggesting that the situation will not improve.

Table 2. Summary of Measures of Capital Replenishment

	Average Age (Years)	Renewal Period (Years)
Major Industries	12	14
Aerospace Firms	15	18
Federal Government	16	22
DoD T&E Sites	34	64

These findings for the DoD T&E sites as a whole apply to each service individually: The Army, Navy and Air Force T&E sites all have longer capital renewal periods and older capital stock than the other sectors. In fact, the services are in much the same position, their capital stock having roughly the same average age.

Our study is too broad to answer such questions as whether the T&E sites are buying the right kinds of capital, whether capital and labor inputs are being kept in balance, and, the ultimate question, whether the DoD sites are maintaining their ability to perform specific missions.

To obtain the above findings, the T&E sites and the other sectors were compared graphically, first with regard to the average age of

---

2 Under steady-state conditions (stable investment, no block obsolescence), if we replace machines every 60 years, their ages will vary from one to 60 years and their average age will be half as long, or 30 years. Put differently, the renewal period is the *maximum* age, which is twice the *average* age under stable conditions. (This relationship between renewal period and average age does not hold for data presented in Table 3 because steady-state conditions did not necessarily apply.)

capital and then using the capital renewal period. For each measure, we considered three separate cases. The first used time averages in order to compare the sectors at the most aggregate level.[3] This was followed by a more detailed comparison to see if the findings hold separately for equipment and structures. (Structures are mainly shells for equipment, and the main penalty for using old buildings is high maintenance cost. Old equipment imposes an added burden in that it may be outmoded and unable to keep up with the demands of advancing technology.) Finally, we examined trends in order to see if the findings are stable over time.

DATA

This section describes the data in Table 3, which were used in the graphical comparisons presented in Section III and in the summary in Table 2. The yearly data used in the trend graphs are voluminous and are therefore not shown. All figures have been converted to constant FY 1988 dollars (using deflators to be described).

DEFINITIONS OF SECTORS

The Major Industry Classes shown in Table 3 are standard categories for Department of Commerce data. These industries are huge, inhomogeneous groupings, and none are good civilian analogues of military T&E sites in terms of the type of capital they use or the kind of activity they pursue. (Table A-1 in Appendix A provides a more detailed breakdown, but these classifications are also huge.)

We tried to obtain data on a smaller industry classification that was close to military T&E, but failed. The closest we got were figures for a three-digit industry called Miscellaneous Business Services (SIC 739 in the 1972 Standard Industrial Classification Manual

---

[3] As stated above, we used the renewal period as an indicator of current capital replacement policies. Using an historical time-average for the renewal period frustrates this interpretation somewhat, but gives us a more stable indicator. Moreover, our trend analysis (described later) shows that the average renewal periods for the aggregate public and private sectors are typical of the figures calculated for the most recent years.

Table 3. Measures of Capital Replenishment. (Average Over Time)

	Capital Stock ($B FY88)	Yearly Investment ($B FY88)	Average Age of Capital (Years)			Capital Renewal Period (Years)		
			Equipment	Structures	Total	Equipment	Structures	Total
Major Industry Classes[a]								
I. Agriculture, Forestry and Fisheries	326.6	23.2	6.1	15.4	10.1	11.3	26.8	14.9
II. Mining	434.6	37.5	5.7	8.6	8.0	10.6	13.7	12.4
III. Construction	101.0	10.0	5.3	8.9	5.9	11.0	14.2	11.4
IV. Manufacturing	1,141.5	79.2	8.6	13.1	10.1	12.3	21.9	14.4
V. Transportation and Public Utilities	1,564.3	86.3	9.3	20.2	15.6	12.1	28.4	18.2
VI. Wholesale Trade	142.1	16.2	5.1	11.1	7.5	7.3	14.2	9.0
VII. Retail Trade	258.1	21.7	5.5	12.4	9.6	8.5	16.4	11.9
VIII. Finance, Insurance and Real Estate	717.5	51.3	5.1	14.5	12.2	7.1	21.8	14.3
IX. Services	360.1	33.2	5.3	12.4	8.5	7.9	20.1	10.9
Total	5,045.9	358.7	7.5	15.1	11.7	10.3	20.2	14.0
Nine Aerospace Firms[b]	4.12	0.33			15.0			18.0
Federal Government[a]								
Non-Military	412.2	14.0	9.9	20.2	17.9	19.3	36.7	30.1
Military	648.1	36.2	9.4	24.8	14.7	13.8	80.0	19.4
Total	1,060.3	50.3	9.5	22.1	15.9	14.4	46.9	22.1
DoD Test and Evaluation Sites[c]								
Army	5.1	0.085			38.1			59.6
Navy	3.6	0.093			37.4			38.2
Air Force	14.8	0.188			31.3			78.6
Total DoD	23.4	0.366			33.7			63.9

[a] All figures are averages for 1965-85.

[b] All figures are averages for 1970-83, but excluding 1972 when there was negative investment.

[c] Capital Stock is for 1985, Yearly Investment is an average for 1981-86, and Capital Renewal Period is the ratio of these two figures. Average Age is for 1987.

published by the Office of Management and Budget), which includes Research and Development Laboratories and Commercial Testing Laboratories. These laboratories, however, are not the ones associated with manufacturing companies, which are the most similar to the military T&E sites. The manufacturing laboratories are reported as part of the manufacturing establishments themselves. To make matters worse, SIC 739 includes businesses that have nothing to do with testing—photofinishing, interior designing, sign painting, telephone message services, etc.

The nine aerospace firms, treated in the aggregate to protect their proprietary interests, are:

- General Dynamics – Fort Worth Division
- General Electric Aircraft Engine Business Group
- Grumman Aerospace Corporation
- Lockheed California Company
- McDonnell Aircraft Company
- Northrop Aircraft Division
- Pratt & Whitney Aircraft Group
- Sikorsky Aircraft Division
- Vought Texas.

The federal government is divided into non-military and military components, and the DoD Test and Evaluation Sites are the 22 bases of the MRTFB shown in Table 1.

The first two columns in Table 3 give the average values of capital stock and investment. These figures are shown to give an idea of the relative sizes of the various sectors. The capital renewal period used in the graphical comparisons and shown in the right-hand columns of Table 3 were obtained by dividing capital stock by investment for each year, and then averaging these yearly renewal periods over time. The values of average age used in the graphical comparisons, also shown in Table 3, were obtained directly from the data except in the case of the aerospace firms, for which data were lacking. Here, we estimated average age by multiplying the renewal period by the ratio of age to renewal period for the major industries.

The remainder of this section states the guidelines used in compiling the data and gives the sources of the data on capital stock, investment, and age.

DATA GUIDELINES

We used several guidelines in compiling the data so that they would be consistent with the figures obtained for the T&E sites:

- Capital is measured by undepreciated (gross) capital stock, which equals total past purchases of capital reduced by discards—equipment or buildings trashed or sold to other firms. (Net capital stock allows for depreciation as well; Appendix B estimates the effect of using gross, rather than net capital.)
- Undepreciated (gross) investment was also used.
- The measures of capital and investment include only equipment and structures, not land.

Exceptions to these guidelines are noted in the following subsection.

DATA SOURCES

Major Industries and Federal Government

The data for the industry and government sectors were obtained from the Bureau of Economic Analysis [1] and escalated to 1988 dollars using the Bureau of Labor Statistics producer price index for "Finished Goods, Capital Equipment" [2].

Aerospace Firms

Capital stock and investment for the aerospace firms were constructed from accounting data obtained by IDA for use in a series of studies on indirect costs of defense contractors. We were not able to completely eliminate depreciation from the capital stock, which we constructed by adding yearly gross investment to a depreciated book value provided by the firm for an early base year (1965-70 for the various firms). Also, expenditures on land included in the figures could not be identified and subtracted from the data.

The accounting data for the aerospace firms did not include figures on the age of capital. We estimated the average age for these firms by multiplying the renewal period by the ratio of average age to renewal period for the industry sector as a whole (18.0 x 11.7/14.0 from Table 3).

T&E Sites

The values of capital stock for the T&E sites were obtained from a separate IDA analysis of the real property inventories maintained by the three services referenced in [3 through 5]. These inventories list the historical expenditures that were funded under the Military Construction (MilCon) appropriation. MilCon generally covers the cost of: (1) the original structures, (2) major equipment such as heating systems and test stands installed at the time of original construction, (3) later additions to structures (such as a new wing), and (4) later additions of major equipment that were procured as part of structural changes funded under MilCon. Appendix B estimates the bias due to neglecting the costs of items not procured under MilCon.

The MilCon expenditures were escalated to a base year by service civil engineers using MilCon deflators. The next steps were to (1) delete military family housing, morale, welfare and recreation, and non-appropriated fund facilities, (2) convert the figures to FY 1988 dollars, and (3) sum the results to get the values of gross capital stock shown in Table 4.

Table 4. Value and Age of Capital Stock at DoD Test and Evaluation Sites

	Capital Stock ($M FY88) by Age Bracket						Total Value of Capital Stock ($M FY88)	Average Age of Capital Stock (Yrs)
	50+ (60)	40-49 (44.5)	30-39 (34.5)	20-29 (24.5)	10-19 (14.5)	<10 (4.5)		
Army	894	1,714	894	1,141	324	126	5,093	38.1
Navy	9	2,089	776	408	154	126	3,562	37.4
Air Force	595	3,456	5,304	2,894	765	1,760	14,774	31.3
Total	1,498	7,259	6,974	4,443	1,243	2,011	23,428	33.7

Note: Numbers in parentheses are the mid-point in years of the age bracket.

The average age of capital was computed by multiplying the capital value by the mid-point year in each age bracket, summing the results, and dividing by the total capital value.

Investment at the T&E sites (Table 5) was obtained by adding up yearly expenditures on Investment and Modernization (I&M) and MilCon obtained from a budget data base being constructed by IDA.

Table 5. Capital Investment Expenditures at DoD Test and Evaluation Sites

	Millions of FY 1988 Dollars						
	1981	1982	1983	1984	1985	1986	Average
Army							
Improvement and Modernization	41.7	58.2	61.6	64.3	76.6	76.0	63.1
Military Construction	11.8	5.9	30.7	20.6	35.3	29.7	22.3
Total	53.5	64.1	92.3	84.9	111.9	105.7	85.4
Navy							
Improvement and Modernization	67.2	92.8	96.8	73.8	70.2	41.6	73.8
Military Construction	27.3	10.8	11.3	50.0	7.7	9.5	19.4
Total	94.5	103.6	108.1	123.8	77.9	51.1	93.2
Air Force							
Improvement and Modernization	50.7	73.8	132.3	204.9	205.9	154.2	137.0
Military Construction	33.0	9.9	85.4	58.8	38.7	79.7	50.9
Total	83.7	83.7	217.7	263.7	244.6	233.9	187.9
DoD							
Improvement and Modernization	159.6	224.8	290.7	343.0	352.7	271.8	273.9
Military Construction	72.1	26.6	127.4	129.4	81.7	118.9	92.6
Total	231.7	251.4	418.1	472.4	434.4	390.7	366.5

Note: Figures are not exact due to rounding errors.

ANALYSIS

The comparisons between the various sectors are shown in six graphs constructed from the data in Table 3. The first three graphs compare the average age of capital, and the second three the renewal period.

Average Age of Capital

Figure 1 shows that the T&E sites are using much older capital than the other sectors: 30-40 years for the T&E sites compared with under 20 years for the other sectors. (The individual industries are labelled as they are in Table 3; industry I, for example, is Agriculture, Forestry and Fisheries.)

Figure 2 distinguishes between equipment and structure in the industry and federal government data (the only data that made the distinction).[4]

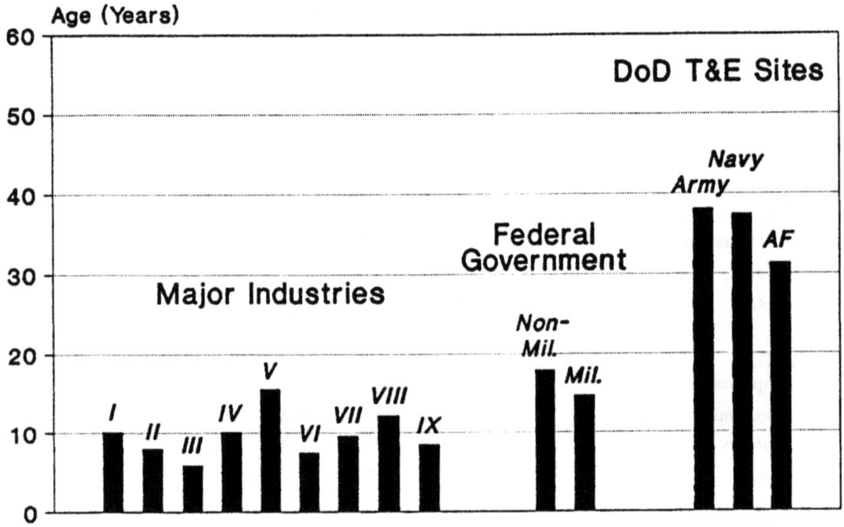

Figure 1. Average Age of Capital

---

4 In the Bureau of Economic Analysis data from which Figure 2 was obtained, "structures" includes major fixed equipment such as heating systems that are installed during initial construction.

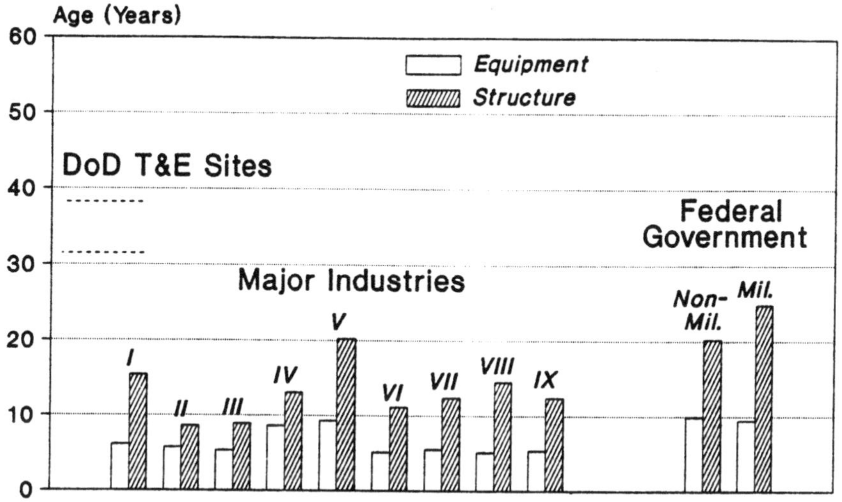

Figure 2. Average Age of Capital: Equipment vs. Structure

The consistent finding is that private and public organizations keep structures longer than they keep equipment. Our data for the T&E sites do not give separate ages for equipment and structure; however, the age of the T&E sites for structure and equipment taken together far exceeds the age of either equipment or structure for the industries or government. (The dotted lines for the DoD T&E sites indicate the range of age for the three military services.)

Figure 3 indicates that the average age of the private sector has been fairly stable over time, suggesting that the T&E sites have been using older capital not only by comparison with the average over time for that sector, but also on a yearly basis. The increasing trend for government structures, however, indicates that these facilities are now somewhat older than their average age over time. For example, the average age for military strucutres is now approximately 30 years, higher than the average figure of 25 years shown in Figure 2.

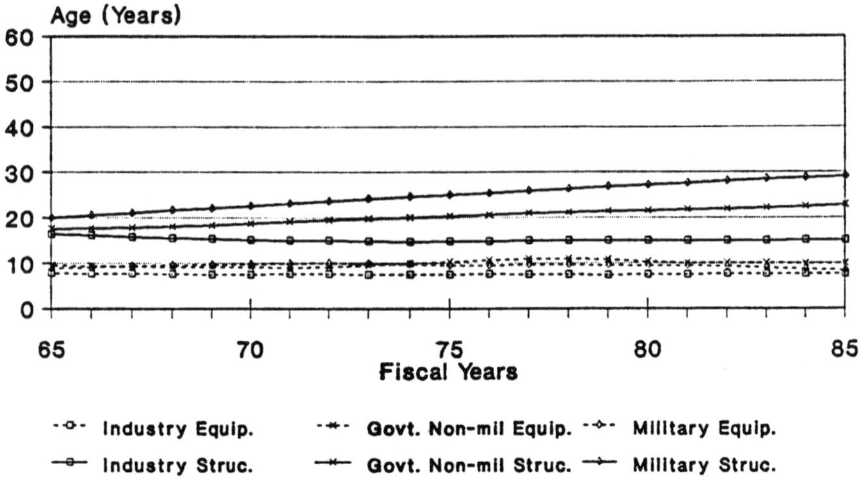

Figure 3. Average Age of Capital: Trends

Capital Renewal Period

Figure 4 shows that the T&E sites renew capital far slower than do other sectors: 40-80 years compared with under 20 years for the industries, aerospace firms, and the military component of the federal government, and about 30 years for the non-military component.

By comparing the renewal periods in Figure 4 with the average ages in Figure 1 (or using the numbers in Table 3), we can predict how the average age will change in the future if current investment rates persist. Under steady-state assumptions, the capital renewal period of 60 and 38 years for the Army and Navy sites will eventually lead to average ages of 30 and 19 years, respectively, well under their current ages of around 38 years. (See footnote 3.)

The Air Force has the lowest average age of the three services at present, but it also has a much longer renewal period. Unless something changes, the renewal period will drive the average age up to 39 years, higher than the current 31 years. Taking the three services together (the "DoD" entry in Table 3), the renewal period is about twice the average age, indicating that the current age of 34 years will persist under present investment levels.

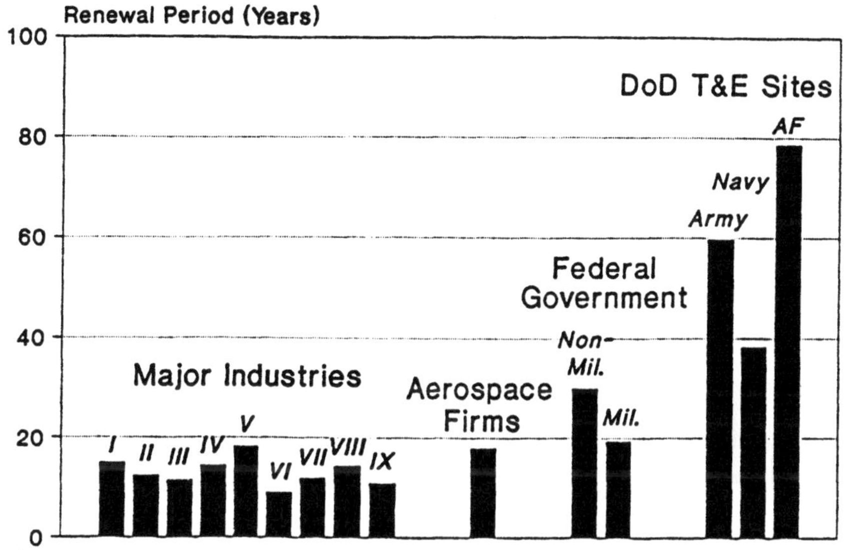

Figure 4. Capital Renewal Period

Figure 5 shows that the T&E sites are doing somewhat better than the structure component of the entire military sector, but all the non-military sectors are renewing capital more aggressively than the T&E sites. The only way that the T&E sites could compare better with the other sectors is if the renewal periods for the T&E sites were the average of a huge structure component and a small equipment component comparable in size to that of the other sectors. However, this is not the case. Measured in value terms, most of the capital at military T&E sites is equipment, not structure.

The trends in Figure 6 show that the long-term renewal periods for the private sector shown in Figures 4 and 5 are typical. For government non-military structures, however, the average renewal period over time of 37 years (Figure 5) is considerably less than its current value of 50 years (Figure 6).

We are left with the compelling finding that the T&E sites are behind the other sectors in terms of the rate at which they are renewing capital.

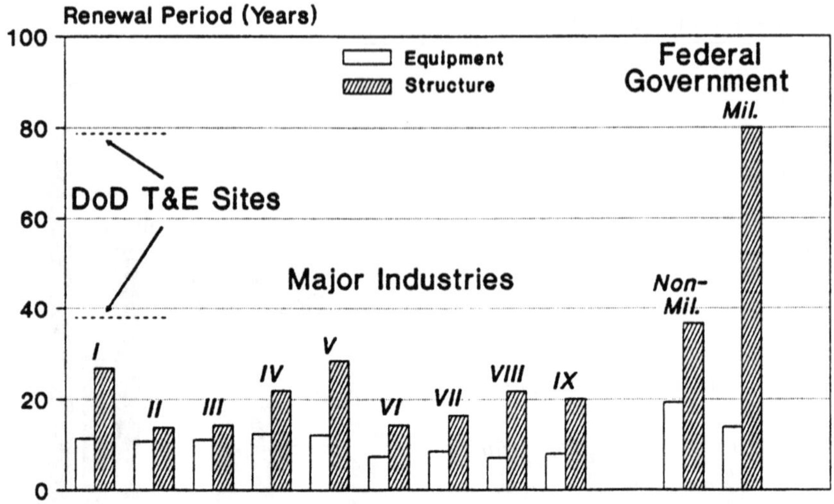

Figure 5. Capital Renewal Period: Equipment vs. Structure

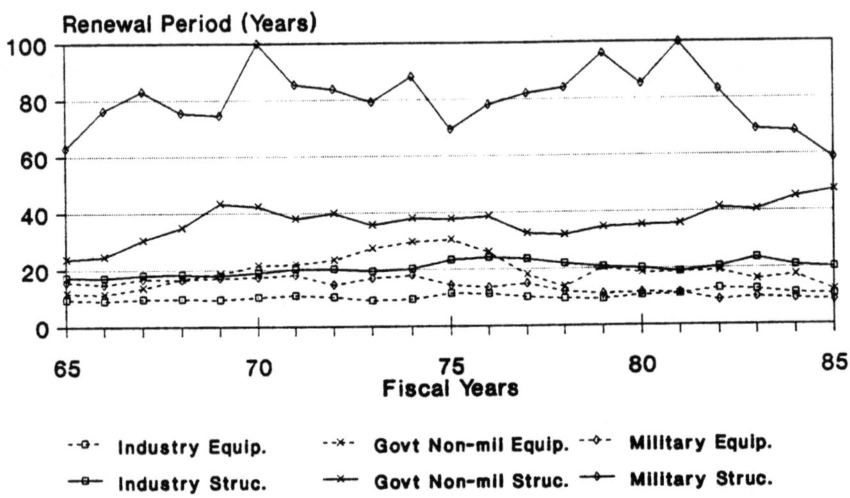

Figure 6. Capital Renewal Period: Trends

APPENDIX A: ADDITIONAL DATA

Table A-1 shows the detailed breakdown of the measures of capital replenishment.

Table A-1. Breakdown of Measures of Capital Replenishment (Averages Over Time)

	Capital Stock ($B FY88)	Yearly Investment ($B FY88)	Average Age of Capital (Years)			Capital Renewal Period (Years)		
			Equipment	Structures	Total	Equipment	Structures	Total
MAJOR INDUSTRY CLASSES[a, b]								
By Type of Output								
I. Agriculture, Forestry and Fisheries	326.6	23.2	6.1	15.4	10.1	11.3	26.8	14.9
II. Mining	434.6	37.5	5.7	8.6	8.0	10.6	13.7	12.4
III. Construction	101.0	10.0	5.3	8.9	5.9	11.0	14.2	11.4
IV. Manufacturing	1,141.5	79.2	8.6	13.1	10.1	12.3	21.9	14.4
Durable goods	616.8	42.9	9.0	12.7	10.2	12.6	21.6	14.4
Machinery, except electrical	92.9	7.1	10.0	11.9	10.6	12.5	17.9	13.6
Electric and electronic equipment	67.2	6.6	7.0	11.1	8.4	9.0	15.9	10.6
Motor vehicles and equipment	75.0	5.8	7.1	12.8	8.3	12.8	27.1	14.1
Nondurable goods	524.7	36.3	8.0	13.5	10.1	12.0	22.4	14.4
V. Transportation and Public Utilities	1,564.3	86.3	9.3	20.2	15.6	12.1	28.4	18.2
Transportation	593.2	25.9	10.4	29.7	20.0	14.7	56.2	23.3
Local and interurban passenger transit	13.7	0.4	9.4	30.5	18.5	22.8	171.0	42.8
Water transportation	51.7	3.2	11.6	20.0	11.7	17.9	42.2	17.9
Transportation by air	60.5	5.4	6.8	10.8	7.0	12.0	20.8	12.3
Transportation services	32.8	1.7	11.3	18.5	11.5	25.8	30.6	25.5
Communication	329.9	27.6	6.5	14.1	10.1	8.8	19.8	11.9
Electric, gas, and sanitary services	641.1	32.8	10.2	15.8	14.2	13.5	24.5	19.7
VI. Wholesale Trade	142.1	16.2	5.1	11.1	7.5	7.3	14.2	9.0
VII. Retail Trade	258.1	21.7	5.5	12.4	9.6	8.5	16.4	11.9
VIII. Finance, Insurance and Real Estate	717.5	51.3	5.1	14.5	12.2	7.1	21.8	14.3

Table A-1. Breakdown of Measures of Capital Replenishment (Averages Over Time) (Continued)

	Capital Stock ($B FY88)	Yearly Investment ($B FY88)	Average Age of Capital (Years)			Capital Renewal Period (Years)		
			Equipment	Structures	Total	Equipment	Structures	Total
IX. Services	360.1	33.2	5.3	12.4	8.5	7.9	20.1	10.9
Auto repair, services, and garages	58.3	6.5	5.6	12.8	6.4	8.5	20.3	9.2
Total	5,045.9	358.7	7.5	15.1	11.7	10.3	20.2	14.0
By Type of Equipment								
Information processing and related equipment	306.9	48.0	5.4			7.3		
Office, computing and accounting machinery	46.6	13.7	4.2			5.5		
Communications equipment	172.9	20.8	6.0			8.3		
Instruments, photocopy and related equipment	87.3	13.5	4.7			6.4		
Industrial equipment	938.0	67.3	8.9			14.5		
Fabricated metal products	101.8	9.6	7.6			10.8		
Engines and turbines	58.8	3.3	11.5			20.1		
Metalworking machinery	205.5	15.7	8.8			13.2		
Special industry machinery, n.e.c.	190.7	13.2	8.5			14.6		
General industrial, including materials handling	189.1	14.7	5.8			12.9		
Electrical trans., distribution, and industrial apparatus	192.0	10.8	12.3			17.6		
Transportation and related equipment	541.6	48.2	7.6			16.7		
Trucks, buses, and truck trailers	183.2	24.1	4.6			7.7		
Autos	86.2	7.1	3.1			13.5		
Aircraft	87.3	8.6	6.5			10.5		
Ships and boats	56.9	3.4	11.8			18.4		
Railroad equipment	128.0	4.9	13.8			35.2		

Table A-1. Breakdown of Measures of Capital Replenishment (Averages Over Time) (Continued)

	Capital Stock ($B FY88)	Yearly Investment ($B FY88)	Average Age of Capital (Years): Equipment	Average Age of Capital (Years): Structures	Average Age of Capital (Years): Total	Capital Renewal Period (Years): Equipment	Capital Renewal Period (Years): Structures	Capital Renewal Period (Years): Total
By Type of Structure								
Industrial buildings	390.0	18.3		12.9			22.7	
Commercial buildings	570.2	34.6		12.5			16.7	
Electric light and power	317.5	14.4		15.4			22.3	
NINE AEROSPACE FIRMS[c]	4.12	0.33			15.0			18.0
FEDERAL GOVERNMENT[b]								
Non-Military	412.2	14.0	9.9	20.2	17.9	19.3	36.7	30.1
Military	648.1	36.2	9.4	24.8	14.7	13.8	80.0	19.4
Total	1,060.3	50.3	9.5	22.1	15.9	14.4	46.9	22.1
By Type of Equipment								
Non-Military	92.2	5.2			9.9	19.3		
Military	424.1	33.4			9.4	13.8		
By Type of Structure[a]								
Non-Military	319.9	8.88			20.2		36.7	
Industrial buildings	61.3	1.36			18.2		51.8	
Educational buildings	1.7	0.05			112.3		63.1	
Other buildings[d]	35.6	1.10			18.6		41.4	
Other structures[e]	9.4	0.33			19.5		39.0	
Military	224.0	2.85			24.8		80.0	
DoD TEST AND EVALUATION SITES[f]								
Army	5.1	0.085			38.1			59.6
Navy	3.6	0.093			37.4			38.2
Air Force	14.8	0.188			31.3			78.6
Total DoD	23.4	0.366			33.7			63.9

[a] For some industries, only selected categories are shown. For example, Lumber and Wood Products are not shown under Durable Goods. Thus, values of Capital Stock and Yearly Investment for subcategories do not always sum to the values for the major industry.

[b] All figures are averages for 1965-85.

[c] All figures are averages for 1970-83, but excluding 1972 when there was negative investment.

[d] Office buildings, police and fire stations, courthouses, garages, passenger terminals, etc.

[e] Electric and gas facilities, transit systems, airfields, etc.

[f] Capital Stock is for 1985, Yearly Investment is an average for 1981-86, and Capital Renewal Period is the ratio of these two figures. Average Age is for 1985.

APPENDIX B: DATA BIASES

This appendix discusses the major biases in the data.

EXCLUSION OF NON-MILCON PURCHASES

In our earlier discussion of the data for the T&E sites in, we noted that our measure of capital stock excludes capital investments that were not made under the MilCon appropriation. This means that age and renewal period have been miscalculated. We estimated the effect of removing this bias by taking our figures for Improvement and Modernization (I&M), the budget category that the T&E sites use for non-MilCon investments, and adding these expenditures to our previous capital stock measure.

Table 5 in the main text shows that the DoD T&E sites have been spending about $274 million yearly for I&M. We added another $50 million yearly as an upper bound of current estimates of unreported I&M—expenditures on equipment that program managers occasionally purchase to test their systems, and that are not entered into base financial records. The result was $324 million yearly. Accumulating this for 34 years (the average age of the capital stock, from Table 4) added an additional $11,000 million to our estimate of capital stock. This means a 47-percent increase in the current capital stock of $23,400 million (Table 4), which would lead to a similar percentage increase in our estimate of capital renewal period, raising it from the current 64 years (Table 3) to over 90 years!

The $50 million of unreported I&M, and the assumption that DoD has been spending $324 million for I&M in past years may both be overestimates, but the calculations do suggest that 64-year capital renewal rate may be too low. If it is low, the future expectation for average age of the capital stock could substantially exceed the current 32 years. On the other hand, the figures for average age calculated in the text are probably overestimated because non-MilCon purchases were excluded. These purchases are generally for equipment bought for existing structures, equipment that may have been bought decades later.

USE OF GROSS CAPITAL STOCK

Because the T&E sites employ older capital, our use of gross rather than net capital stock led to overestimations of the age and renewal period of the T&E sites relative to the other sectors. We obtained estimates of age and renewal period using net figures by calculating net/gross ratios for age and renewal period for a simple steady-state case and applying these ratios to the gross figures listed in Table 3 of the main text. For these calculations, we assumed an annual depreciation of two percent yearly to account for declining productivity relative to advancing technology. (The use of gross, or undepreciated capital assumes that as long as the capital is in use, it is as productive as new capital costing the same amount in constant dollars. The rapid advances in avionics and other defense technologies make this an unrealistic assumption.)

Table B-1 illustrates how the net/gross ratios were calculated, using a simple case in which we invest one dollar yearly in capital with a 4-year lifetime. (Capital bought prior to 1985 is no longer available, by definition.)

Table B-1. Gross vs. Net Capital Stock Measures

			Capital Stock at End of 1988	
Year	Investment at Start of Year	Age at End of 1988	Gross	Net
1985	1	4	1	.9412(.98^3)
1986	1	3	1	.9604(.98^2)
1987	1	2	1	.98
1988	1	1	1	1

The average age of the capital stock is the sum of the age of each vintage of capital weighted by the respective capital stock (gross or net); the renewal period equals the capital stock (gross or net) summed over years, divided by the annual investment of one dollar, as shown below:

	Gross	Net	Net/Gross Ratio
Average Age	2.5	2.475	0.990
Renewal Period	4.0	3.882	0.971

This technique was used to calculate net/gross ratios for the various sectors. (Capital lifetimes were simply estimated to be twice the average ages of capital shown in Table 3.) These ratios and their application to the gross capital figures from Table 3 are shown in Table B-2. The T&E sites are now closer to the other sectors, but there is still a sizable gap. And the improvement would be counteracted by the MilCon adjustments discussed in the previous section.

Table B-2. Net Average Age and Renewal Period

	Average Age			Renewal Period		
Sector	Gross	Ratio	Net	Gross	Ratio	Net
Industry	11.7	0.74	8.7	14.0	0.80	11.2
Aerospace	15.0	0.73	10.9	18.0	0.79	14.2
Government	15.9	0.67	10.7	22.1	0.75	16.6
T&E Sites	33.7	0.47	15.8	63.9	0.59	37.7

LIMITATIONS OF THE SERVICES' REAL PROPERTY INVENTORIES

Most real property inventories maintained by the services allow only a single "year" entry for all expenditures that are made during a building's life, and it is usually the year of original construction.[5]

5 Those who maintain the inventories occasionally try to compensate for the error by listing especially large expenditures *and all previous ones* in the year of the most recent one. This practice is not systematic, however, and probably done with too little frequency to make much of a difference.

This practice tends to overstate capital stock (and hence renewal period) because those expenditures that are made after initial construction are over-inflated. There is an opposite bias in that some expenditures are not entered at all, but we do not know the magnitude, and our measures of capital stock and renewal period could be off in either direction.

Entering building modifications at the age of original construction does, however, impart an unambiguous bias to average age: it is over-stated. We did not estimate the extent, but the fact that the average age we calculated is about half the renewal period suggests that the bias is not large.

REFERENCES

[1] U. S. Department of Commerce, Bureau of Economic Analysis. *Fixed Reproducible Tangible Wealth in the United States, 1925-85*. Washington, DC: U.S. Government Printing Office, June 1987.

[2] U.S. Bureau of the Census. *Statistical Abstract of the United States: 1987* (107th edition.) Washington, DC: U.S. Government Printing Office, December 1986.

[3] Department of the Army summary facility data as of September 1987, provided by the RPMA Branch, Facilities Division, Office of the Chief of Engineers (DAEN-ZCF-R).

[4] "P-164, Detailed Inventory of Naval Shore Facilities, 30 Sept 1987," provided by the Planning Support Branch, Facilities Planning Division, Naval Facilities Engineering Command.

[5] "Air Force Real Property Inventory Detail List as of 30 Sept 1987 (HAFLEE 7115)," provided by the Real Property Division, Office of the Deputy Chief of Staff, Logistics and Engineering, Headquarters Air Force (AF/LEER).

MIX
Papier aus verantwortungsvollen Quellen
Paper from responsible sources
FSC® C105338

If you have any concerns about our products,
you can contact us on
ProductSafety@springernature.com

In case Publisher is established outside the EU,
the EU authorized representative is:
Springer Nature Customer Service Center GmbH
Europaplatz 3, 69115 Heidelberg, Germany

Printed by Libri Plureos GmbH
in Hamburg, Germany